Mathematical Theory
of Reliability

The SIAM Series in Applied Mathematics
R. F. Drenick, Harry Hochstadt, Dean Gillette, *Editors*

RIORDAN, *Stochastic Service Systems*
ROSENBLATT, *Time Series Analysis*
HENRICI, *Error Propagation for Difference Methods*
MILLER, *Multidimensional Gaussian Distributions*
BARLOW AND PROSCHAN, *Mathematical Theory of Reliability*

In preparation (*titles tentative*):
BAREISS, *Transport Theory*
DORN AND LEMKE, *Mathematical Programming*
HEADING, *The Stokes Phenomenon*
ISAACS, *Differential Games*
KELLEY, *Critical-Path Scheduling*
LEE AND MARKUS, *Foundations of Optimum Control Theory*
MACCOLL, *Modern Dynamical Theory*
MARSAGLIA, *Monte Carlo Methods*
NOETHER, *Nonparametric Statistics*
SNEDDON AND LOWENGRUB, *Crack Problems*

Mathematical Theory of Reliability

Richard E. Barlow
University of California at Berkeley,

Frank Proschan
Boeing Scientific Research Laboratories
Seattle, Washington

with contributions by Larry C. Hunter
Sylvania Electronic Defense Laboratories,
Mountain View, California

John Wiley & Sons, Inc., New York · London · Sydney

10 9 8 7 6 5

Copyright © 1965 by John Wiley & Sons, Inc.

All Rights Reserved

This book or any part thereof
must not be reproduced in any form
without the written permission of the publisher.

ISBN 0 471 04965 4
Library of Congress Catalog Card Number: 64-25897
Printed in the United States of America

Preface

What is mathematical reliability theory? Generally speaking, it is a body of ideas, mathematical models, and methods directed toward the solution of problems in predicting, estimating, or optimizing the probability of survival, mean life, or, more generally, life distribution of components or systems; other problems considered in reliability theory are those involving the probability of proper functioning of the system at either a specified or an arbitrary time, or the proportion of time the system is functioning properly. In a large class of reliability situations, maintenance, such as replacement, repair, or inspection, may be performed, so that the solution of the reliability problem may influence decisions concerning maintenance policies to be followed.

Since reliability theory is mainly concerned with probabilities, mean values, probability distributions, etc., it might be argued that the theory is simply an application of standard probability theory and really deserves no special treatment. This view would be as shortsighted as arguing that probability theory itself is simply an application of standard mathematical theory and deserves no special development. Reliability problems have a structure of their own and have stimulated the development of new areas in probability theory itself. We shall see this most clearly in developing and applying the concept of monotone failure rates and in obtaining new results in renewal theory as a result of comparing competing replacement policies.

The monograph presents a survey of some mathematical models useful in solving reliability problems. The models are *probabilistic*; that is, from information concerning the underlying distributions, such as for component lives, deductions are made concerning system life, optimum system design, etc. *Statistical* problems, such as the estimation of component or system life from observed data, are generally not treated. A larger and possibly more useful volume could easily be devoted to statistical problems in reliability. We have, however, chosen to confine ourselves to the more developed subject of probabilistic models in reliability theory.

Throughout the book the emphasis is on making minimal assumptions, and only those based on plausible physical considerations, so that the resulting mathematical deductions may be safely made about a large variety of commonly occurring reliability situations. Thus quite often we assume only that the life distributions have an increasing (decreasing) failure rate and that only a single parameter, such as the mean, is actually known. (Throughout the book we write "increasing" in place of "nondecreasing" and "decreasing" in place of "nonincreasing.") From these modest assumptions a surprising number of useful deductions can be made. This theme of monotone failure rate is one of the unifying influences in our attempt to integrate the existing reliability theory.

In selecting material we must admit we have favored reliability models we have worked on ourselves. This method of selection may not be completely disastrous, for in originally turning our attention to these particular models we were influenced by their importance to reliability theory and by their mathematical interest. We are painfully aware of the resulting gaps and omissions, however. If the omissions are too egregious, we hope an irate reader will write a monograph completing the picture!

To whom is the book directed? As is true of all the books in this series, this monograph is addressed primarily to the applied mathematician. We hope the monograph will be of interest not only to those already working on reliability problems, but also to scientists and engineers who are working in other areas but are interested in seeing some applications of probability theory in new areas. The mathematical subject chiefly used in the monograph is probability theory and, to a limited degree, the concept of total positivity. An appendix is provided to develop the theory of total positivity required.

In the introductory chapter we present a historical survey of the subject; such a survey we feel is greatly needed, for in the development of reliability theory, research has all too often been undertaken and published without an awareness or use of what has gone on before. The chapter is completed by constructing a single general mathematical model which by appropriate specialization yields the basic quantities of interest in reliability theory. Chapter 2 surveys probability distributions of interest in reliability. The point of view is to develop the properties not just of specific distributions, but of broad classes of distributions characterized by plausible physical assumptions. This point of view motivates the development of the

PREFACE

properties of the class of distributions characterized by an increasing failure rate, corresponding to the physical phenomenon of wear-out. From this modest assumption applying in so many actual situations, an amazingly rich set of consequences ensues.

Chapters 3 and 4 are devoted to maintenance models; Chapter 3 is concerned with the stochastic aspects of various maintenance policies and Chapter 4 with the optimization of these policies. In Chapter 3 an unexpected bonus is obtained; by comparing two maintenance policies, some useful bounds on the renewal function are obtained. For distributions with increasing failure rate with only the mean known, the use of the bounds permits determination of the renewal function within an error of one-half. In Chapter 4 optimum stationary and time-dependent policies are obtained or their characteristics described for a number of classes of replacement, repair, and inspection models. The assumption of an increasing (decreasing) failure rate will be seen to play a crucial role in simplifying the solution and in leading to useful qualitative results concerning the form of the optimum policy.

Chapter 5 presents stochastic models for complex systems. The key probability tools are Markov and semi-Markov processes. By using these tools, optimal control rules are developed, repairman-type problems solved, and monitoring and marginal testing policies analyzed.

Chapters 6 and 7 are devoted to models involving redundancy. Chapter 7 shows how to allocate redundancy among the various subsystems under linear constraints on weight, volume, cost, etc., so as to maximize system reliability. The model is applicable in both system design and spares provisioning. Chapter 7 develops qualitative relationships between system reliability and component reliabilities. As an example, conditions are presented under which system life possesses an increasing failure rate when each component life possesses an increasing failure rate.

Finally, appendices are provided summarizing special topics needed in the development of the theory. Thus an appendix on total positivity is presented and shows, among other things, how the concept of an increasing failure rate falls into the more general framework of total positivity. Because the assumption of increasing failure rate is so basic in much of the theory, a statistical test is presented in a second appendix for determining from a sample of observations whether the underlying population does have an increasing failure rate. This test is the one exception to the rule followed throughout the book of confining attention to probabilistic questions rather than statistical ones. Tables are presented in the final appendix giving bounds on survival probability from a knowledge of either one or two moments of an increasing (decreasing) failure rate distribution.

Equation numbers are governed by the section in which the equations appear. Thus Equation (2.3) is the third equation of Section 2. The chapter number is not specified in referring to an equation within the same chapter. In referring to an equation in a different chapter, the number of the chapter is specified. Thus in Chapter 5 we might say "by Equation (2.3) of Chapter 2," whereas in Chapter 2 reference to the same equation would appear as "by Equation (2.3)." Theorem numbers are governed by the same rules. A list of references appears in the back of the book. References within the text include the author's name and the year of publication.

September 1964
Richard E. Barlow
Frank Proschan

Acknowledgments

First, we should like to acknowledge our indebtedness to Larry C. Hunter, who wrote an earlier version of Chapter 4, offered helpful suggestions and advice concerning the remaining chapters, and originally developed with us many of the mathematical models described in the monograph. Second, we want to acknowledge the helpful comments and suggestions of Stuart Bessler, R. F. Drenick, James Esary, Betty Flehinger, Samuel Karlin, Albert Marshall, Ronald Pyke, Judah Rosenblatt, Joan Rosenblatt, I. R. Savage, and George Weiss. Third, we want to thank the Boeing Scientific Research Laboratories, the General Telephone and Electronics Laboratories, Inc., and the Sylvania Electronic Defense Laboratories for their generous support. Finally, we want to thank Mrs. Helene Turner for her patience and skill in typing the manuscript.

R. E. B.
F. P.

Contents

Chapter 1 *Introduction*   *1*

1. Historical Background of the Mathematical Theory of Reliability, 1
2. Definitions of Reliability, 5

Chapter 2 *Failure Distributions*   *9*

1. Introduction, 9
2. Typical Failure Laws, 12
3. The Exponential as the Failure Law of Complex Equipment, 18
4. Monotone Failure Rates, 22
5. Preservation of Monotone Failure Rate, 35
6. Additional Inequalities, 39
7. General Failure Rates, 41

Chapter 3 *Operating Characteristics of Maintenance Policies*   *46*

1. Introduction, 46
2. Renewal Theory, 48
3. Replacement Based on Age, 61
4. Comparison of Age and Block Replacement Policies, 67
5. Random Replacement, 72
6. Repair of a Single Unit, 74

Chapter 4 Optimum Maintenance Policies 84

1. Introduction, 84
2. Replacement Policies, 85
3. Inspection Policies, 107

Chapter 5 Stochastic Models for Complex Systems 119

1. Introduction, 119
2. Markov Chains and Semi-Markov Processes, 121
3. Repairman Problems, 139
4. Marginal Checking, 151
5. Optimal Maintenance Policies under Markovian Deterioration, 156

Chapter 6 Redundancy Optimization 162

1. Introduction, 162
2. Optimal Allocation of Redundancy Subject to Constraints, 163
3. Application to Parallel Redundancy Model, 170
4. Application to Standby Redundancy Model, 175
5. Complete Families of Undominated Allocations, 180
6. Optimal Redundancy Assuming Two Types of Failure, 185

Chapter 7 Qualitative Relationships for Multicomponent Structures 196

1. Introduction, 196
2. Achieving Reliable Relay Circuits, 197
3. Monotonic Structures, 202
4. S-shaped Reliability Functions for Monotonic Structures, 209
5. k-out-of-n Structures, 216
6. Relationship between Structure Failure Rate and Component Failure Rates, 219

Appendix 1 *Total Positivity* *227*

Appendix 2 *Test for Increasing Failure Rate* *232*

Appendix 3 *Tables Giving Bounds on Distributions with Monotone Failure Rate* *236*

References *241*

Index *249*

CHAPTER 1

Introduction

1. HISTORICAL BACKGROUND OF THE MATHEMATICAL THEORY OF RELIABILITY

This is a necessarily incomplete but, we hope, representative survey of the events, people, and literature which have come to be associated with what is now called the mathematical theory of reliability. Some of the areas of reliability research which we document are life testing, structural reliability (including redundancy considerations), machine maintenance problems (a part of queueing theory), and replacement problems (closely connected to renewal theory). Overlapping areas not treated are quality control, extreme value theory, order statistics, and censorship in sampling.

The mathematical theory of reliability has grown out of the demands of modern technology and particularly out of the experiences in World War II with complex military systems. One of the first areas of reliability to be approached with any mathematical sophistication was the area of machine maintenance (Khintchine, 1932, and C. Palm, 1947). The techniques used to solve these problems grew out of the successful experiences of A. K. Erlang, C. Palm, and others in solving telephone trunking problems. The earliest attempts to justify the Poisson distribution as the input distribution of calls to a telephone trunk also laid the basis for using the exponential as the failure law of complex equipment. Erlang and Palm presented heuristic arguments to support the Poisson as a limiting distribution for telephone calls, and finally Khintchine (1960) and Ososkov (1956) established mathematically rigorous proofs and provided necessary and sufficient conditions. Chapter 2 discusses these results in greater detail.

Applications of renewal theory to replacement problems were discussed as early as 1939 by A. J. Lotka, who also summarized earlier work in this area. N. R. Campbell in 1941 also approached replacement problems using renewal theory techniques. W. Feller is generally credited with developing renewal theory as a mathematical discipline (Feller, 1941, 1949).

In the late 1930's the subject of fatigue life in materials and the related subject of extreme value theory were being studied by W. Weibull (1939), Gumbel (1935), and Epstein (1948), among others. Later Gumbel's book (1958) was to painstakingly supply data to illustrate the use of each of the extreme value (asymptotic) distributions to represent lifetimes. In 1939 Weibull, a professor at the Royal Institute of Technology in Sweden, proposed the distribution named after him as an appropriate distribution to describe the life length of materials. It is interesting that he proposed this distribution, apparently without recognizing it as an extreme value distribution. Considering only the class of failure distributions of the form $F(x) = 1 - e^{-\varphi(x)}$, Weibull (1951) advanced the following argument:

> The only necessary condition this function $\varphi(x)$ has to satisfy is to be a positive non-decreasing function, vanishing at a value x_u, which is not of necessity equal to zero.
>
> The most simple function satisfying this condition is $(x - x_u)^m/x_0$ and thus we put $F(x) = 1 - e^{-(x-x_u)^m/x_0}$. The only merit of this distribution is to be found in the fact that it is the simplest mathematical expression of the appropriate form which satisfied the necessary general conditions. Experience has shown that, in many cases, it fits the observations better than other known distribution functions.

Weibull completes his argument by the following remark: "It is believed that ... the only practical way of progressing is to choose a simple function, test it empirically, and stick to it as long as none better has been found." Later, H. E. Daniels (1945) and Z. W. Birnbaum and S. C. Saunders (1958) were to propose mathematical models justifying the assumption of the normal and gamma families respectively in certain situations.

During the 1940's the major statistical effort on reliability problems was in the area of quality control. We omit discussion of the events in this field and refer the interested reader to the introduction to A. J. Duncan's work (1953).

In the early 1950's certain areas of reliability, especially life testing and electronic and missile reliability problems, started to receive a great deal of attention both from mathematical statisticians and from the engineers in the military-industrial complex. Among the first groups to face up seriously to the problem of tube reliability were the commercial airlines (Carhart, 1953, p. 20). Accordingly, the airlines set up an organization called Aeronautical Radio, Inc. (ARINC) which, among other functions, collected and analyzed defective tubes and returned them to the tube manufacturer. In its years of operation with the airlines, ARINC achieved notable success in improving the reliability of a number of tube types. The ARINC program since 1950 has been focused on military reliability problems.

In December 1950 the Air Force formed an *ad hoc* Group on Reliability of Electronic Equipment to study the whole reliability situation and recommend measures that would increase the reliability of equipment and reduce maintenance. By late 1952 the Department of Defense had established the Advisory Group on Reliability of Electronic Equipment (AGREE). AGREE published its first report on reliability in June of 1957. This report included minimum acceptability limits, requirements for reliability tests, effect of storage on reliability, etc.

In 1951 Epstein and Sobel began work in the field of life testing which was to result in a long stream of important and extremely influential papers. This work marked the beginning of the widespread assumption of the exponential distribution in life-testing research. They justify their choice of the exponential distribution with the following short explanation.

After some consideration and after discussion with electronics experts ... we decided to turn our initial efforts to ordered observations drawn from non-normal distributions. Specifically, we decided to study the case where the characteristic X being investigated has an exponential distribution with a density $f(x; \theta)$ of the form

$$f(x; \theta) = \frac{1}{\theta} e^{-x/\theta} \qquad \theta > 0, x \geq 0$$

(Epstein and Sobel, 1953).

In the missile industry such men as Robert Lusser, Richard R. Carhart, and others, were also active at this time in promoting interest in reliability and stating the problems of most interest to their technology. According to one source,† "The basic definition of reliability (as used by engineers) was first presented by Robert Lusser, then of the R & D Division, Redstone Arsenal, in a small symposium on reliability at Convair in San Diego in 1952." Richard R. Carhart, then of the RAND Corporation, displayed a more scientific, although perhaps less colorful, approach to his subject with the publication of his 1953 report.

† A Manual of Reliability, *Product Engineering*, McGraw-Hill Publishing Company, May 1960.

In 1952 D. J. Davis published a paper presenting failure data and the results of several goodness-of-fit tests for various competing failure distributions. This data seemed to give a distinct edge to the exponential distribution, and for this reason the Davis paper has been widely referred to for support of the assumption of an exponential failure distribution. With the publication of this paper and the Epstein-Sobel paper (1953), the exponential distribution acquired a unique position in life testing. This position became even more secure in 1957 with the AGREE report, which considered little else.

A fundamental reason for the popularity of the exponential distribution and its widespread exploitation in reliability work is that it leads to simple addition of failure rates and makes possible the compilation of design data in a simple form. However, even as early as 1955 serious consideration began to be given to other life distributions. J. H. Kao (1956, 1958), among others, was influential in bringing attention to the Weibull distribution. This interest in the Weibull was to grow ever stronger until it became of major importance in 1959 with the publication of the Zelen-Dannemiller paper pointing out that many life test procedures based on the exponential are not robust.

The problem of missile reliability can be considered in terms of confidence intervals for products of binomial parameters (Buehler, 1953). This problem also received attention by G. P. Steck (1957), J. R. Rosenblatt (1963), and A. Madansky (1958).

The missile reliability problem also motivated Z. W. Birnbaum and his co-workers. In a paper published in 1955 he summarized the problem as follows. "If structural components of a mechanism are mass produced, the strength at failure Y of each single component (equals stress at which this component will fail) may be considered a random variable. The component is installed in an assembly and exposed to a stress which reaches its maximum value X, again a random variable. If $Y < X$, then the component will fail in use." Birnbaum proceeds to show how the probability of this event can be estimated.

The mathematically important paper of Moore and Shannon appeared in 1956. This was concerned with relay network reliability and is discussed in some detail in Chapter 7. Moore and Shannon were stimulated by von Neumann's attempt to describe certain operations of the human brain and the high reliability that has been attained by complex biological organisms. Also in 1956 appeared the reports of G. Weiss, at that time at the Naval Ordnance Laboratory, White Oak, which introduced the use of semi-Markov processes to solve system maintainability problems.

In 1958 an elegant summation of the known mathematical results in renewal theory was presented by Walter Smith. As has been mentioned, this theory has many applications to replacement problems (see Chapter 3). In the same year the first extensive bibliography on life testing was published by William Mendenhall (1958), indicating a coming of age in this field.

Largely motivated by vibration problems encountered in the new commercial jet aircraft, Z. W. Birnbaum and S. C. Saunders in 1958 presented an ingenious statistical model for life lengths of structures under dynamic loading. Their model made it possible to express the probability distribution of life length in terms of the load given as a function of time and of deterioration occurring in time independently of loading. The special case of a constant load or of periodic loading with constant amplitude led them to suggest the gamma distribution for life lengths in certain situations.

In late 1958 Sobel and Tischendorf presented an acceptance sampling plan with new life test objectives. Although not mathematically startling, this approach provided a new basis for writing life test specifications. This sampling plan was based on the exponential distribution and was later extended to other life distributions by S. Gupta and M. Sobel.

Partly motivated by the Birnbaum-Saunders (1958) paper, R. F. Tate in 1959 took up the problem of obtaining minimum variance unbiased estimates of functions of parameters from the gamma, Weibull, and two-parameter exponential distributions. In particular, Tate obtained minimum variance unbiased estimates for the probability that a system or structure would survive a specified time based on n observed times to failure.

In 1959 and 1960 a class of reliability and logistic problems were solved in papers by Black and Proschan, with Proschan using some of the techniques of total positivity. This material is discussed in Chapter 6. Also in 1961 appeared a paper by Birnbaum, Esary, and Saunders which significantly extends the earlier work of E. F. Moore and C. E. Shannon (1956). Their work in this particular area is continuing.

2. DEFINITIONS OF RELIABILITY

In considering various reliability problems, we shall wish to analyze and calculate certain quantities of interest, designated in the literature by a variety of labels: reliability, availability, interval availability, efficiency, effectiveness, etc. Unfortunately, the definitions given in the literature are sometimes unclear and inexact and vary among different writers. We hope, nevertheless, that the reader will experience no real difficulty in this monograph since the specific quantities of interest in each problem will be defined explicitly in terms of probabilities or expected values as the problem is being presented.

Even though we do not believe a comprehensive set of definitions is required at this point for understanding the models to follow, it may be of some value to present a unified treatment of the various concepts and quantities involved in the subject of mathematical reliability. Specifically, we shall define mathematically a single generalized quantity which, when appropriately specialized, will yield most of the fundamental quantities of reliability theory.

To this end we assume a system whose state at time t is described by $\mathbf{X}(t) = (X_1(t), \ldots, X_n(t))$, a vector-valued random variable. For example, $\mathbf{X}(t)$ may be the one-dimensional variable taking on the value 1 corresponding to the functioning state and 0 corresponding to the failed state. Alternately, $\mathbf{X}(t)$ may be a vector of system parameter values, with each component $X_i(t)$ ranging over an interval of real numbers. $\mathbf{X}(t)$, being a random variable, will be governed by a distribution function,

$$F(x_1, \ldots, x_n; t);$$

explicitly, $F(x_1, \ldots, x_n; t)$ equals the probability that

$$X_1(t) \leq x_1, \ldots, X_n(t) \leq x_n.$$

Now corresponding to any state $\mathbf{x} = (x_1, \ldots, x_n)$, there is a *gain*, or payoff, $g(\mathbf{x})$. Thus in the two-state example just given, the gain accruing from being in the functioning state ($x = 1$) might be one unit of value, so that $g(1) = 1$, and the gain from being in the failed state ($x = 0$) might be 0, so that $g(0) = 0$. The expected gain $G(t)$ at time t will be a quantity of interest; it may be calculated from

$$G(t) = Eg(\mathbf{X}(t)) = \int \cdots \int g(x_1, \ldots, x_n) \, dF(x_1, \ldots, x_n; t). \quad (2.1)$$

Finally, we may average the expected gain $G(t)$ over some interval of time, $a \leq t \leq b$, with respect to some weight function $W(t)$ to obtain

$$H(a, b) = \int_a^b G(t) \, dW(t). \quad (2.2)$$

Now we are ready to specialize (2.1) and (2.2) to obtain the various basic quantities arising in reliability theory.

1. "*Reliability* is the probability of a device performing its purpose adequately for the period of time intended under the operating conditions encountered" (Radio-Electronics-Television Manufacturers Association, 1955).

Ordinarily "the period of time intended" is $[0, t]$. Let $X(u) = 1$ if the device is performing adequately at time u, 0 otherwise; we assume that

INTRODUCTION   7

adequate performance at time t implies adequate performance during $[0, t]$. Then from (2.1)

$$G(t) = Eg(X(t)) = P[X(t) = 1] = \text{probability that the device performs adequately over } [0, t].$$

Thus $G(t)$ *is* the reliability of the device as defined above. In general, we shall assume that, unless repair or replacement occurs, adequate performance at time t implies adequate performance during $[0, t]$.

2. *"Pointwise availability:* The probability that the system will be able to operate within the tolerances at a given instant of time" (Hosford, 1960, p. 53).

As before we let $X(t) = 1$ if the system is operating within tolerances at time t, 0 otherwise. Also as before $g(1) = 1$, $g(0) = 0$. Now, however, we do not exclude the possibility of repair or replacement before time t. Then

$$G(t) = Eg(X(t)) = P[X(t) = 1] \text{ is the probability that the system is operating within tolerances at time } t.$$

Thus $G(t)$ now yields pointwise availability at the time point t.

The term "availability" is used by Welker and Horne (1960, p. 42), for the same quantity.

3. *"Interval availability:* The expected fraction of a given interval of time that the system will be able to operate within the tolerances" (Hosford, 1960, p. 53). Repair and/or replacement is permitted.

Suppose the given interval of time is $[a, b]$. Then with X, g defined as in item 2 above and $W(t) = (t - a)/(b - a)$, we compute from (2.2)

$$H(a, b) = \frac{1}{b - a} \int_a^b G(t)\, dt = \frac{1}{b - a} \int_a^b Eg(X(t))\, dt,$$

so that under suitable regularity conditions

$$H(a, b) = E \frac{\int_a^b g(X(t))\, dt}{b - a}$$

which is the expected fraction of the time interval $[a, b]$ that the system is operating within tolerances. Thus $H(a, b)$ is the interval availability for the interval $[a, b]$.

Barlow and Hunter (1960a, pp. 46–47) refer to essentially the same quantity as "efficiency."

4. *"Limiting interval availability"* is the expected fraction of time in the long run that the system operates satisfactorily.

To obtain limiting interval availability simply compute $\lim_{T\to\infty} H(0, T)$ in item 3. Barlow and Hunter (1960a, p. 47) call this quantity "limiting efficiency."

5. "*Interval reliability* is the probability that at a specified time, the system is operating and will continue to operate for an interval of duration, say x" (Barlow and Hunter, 1961, pp. 206–207). Repair and/or replacement is permitted. The continued operation during the interval is, of course, to be achieved without the benefit of repair or replacement.

To obtain this quantity, let $X(t) = 1$ if the system is operating at time t, 0 otherwise. Then the interval reliability $R(x, T)$ for an interval of duration x starting at time T is given by

$$R(x, T) = P[X(t) = 1, T \le t \le T + x]. \tag{2.3}$$

Limiting interval reliability is simply the limit of $R(x, T)$ as $T \to \infty$. Truelove (1961, p. 27) calls this "strategic reliability."

Drenick (1960a) formulates a somewhat different general model from which, by appropriate specialization, he derives definitions of certain quantities of interest in reliability. His formulation is in terms of a renewal process. Failures during $[0, t]$ occur at times $t_1, t_2, \ldots, t_n$, $t_1 < t_2 < \cdots < t_n < t$; replacement is made immediately following failure. A gain function $W_n(\mathbf{t}, t)$ describes the economic gain accruing from this outcome, where $\mathbf{t} = (t_1, t_2, \ldots, t_n)$. Thus the expected gain $U(t)$ up to time t is given by

$$U(t) = \sum_{n=0}^{\infty} \int W_n(\mathbf{t}, t) f(\mathbf{t} \mid n, t) \, dt P[N(t) = n] \tag{2.4}$$

where $f(\mathbf{t} \mid n, t)$ is the joint conditional density of failure at times $t_1 < t_2 < \cdots < t_n$ given that $N(t) = n$, $N(t)$ being the number of failures in $[0, t]$. Special cases of (2.4) yield quantities that Drenick calls replacement rate, maintenance ratio, mission success ratio, and mission survival probability.

CHAPTER 2

Failure Distributions

1. INTRODUCTION

A failure distribution represents an attempt to describe mathematically the length of life of a material, a structure, or a device. The modes of possible failure for the item in question will affect the analytic form of the failure distribution. Materials and structures can fail in several ways, with two or more types of failure sometimes occurring at once. Examples of the different types of failure are: static failure when fracture occurs during a single load application; instability of a structure caused by strain energy stored in a member; chemical corrosion such as hydrogen embrittlement; fatigue due to cyclic loading; and sticking of mechanical assemblies. Electronic devices may fail when certain critical parameters drift out of tolerance bounds with changes in time, temperature, humidity, and altitude. Early failures result from improper design, improper manufacture, and improper use (environment). Unfortunately, the choice of a failure distribution on the basis of these physical considerations is still largely an art. In some cases, however, the relationship between the failure mechanism and the failure rate function may be used in making a choice.

On the basis of actual observations of times to failure, it is difficult to distinguish among the various nonsymmetrical probability functions. Thus the differences among the gamma, Weibull, and log normal distribution functions become significant only in the tails of the distribution, but actual observations are sparse in the tails because of limited sample sizes. In order to discriminate among probability functions that cannot be distinguished from each other within the range of actual observation, it is necessary to appeal to a concept that permits us to base the differentiation among distribution functions on physical considerations. Such a concept is based on the failure rate function.

If the failure distribution F has a density f, the *failure rate function* $r(t)$ is defined for those values of t for which $\bar{F}(t) < 1$ by

$$r(t) = f(t)/\bar{F}(t),$$

where throughout the book $\bar{F}(t) = 1 - F(t)$. This function has a useful probabilistic interpretation; namely, $r(t)\,dt$ represents the probability that an object of age t will fail in the interval $[t,\, t + dt]$. It is important in a number of applications and is known by a variety of names. It is used by actuaries under the name of "force of mortality" to compute mortality tables (Steffenson, 1930). In statistics its reciprocal for the normal distribution is known as "Mills' ratio." It plays an important role in determining the form of extreme value distributions, and in extreme value theory is called the "intensity function" (Gumbel, 1958). In reliability theory it has been called "hazard rate."

In many applications there is every reason to believe that beyond a certain age the function $r(t)$ does not decrease because of the inevitable deterioration which occurs. This hypothesis becomes even more credible for devices in which "early failures" have been weeded out by such expedients as the "burn-in test" for vacuum tubes. There are, of course, cases when this function is initially decreasing. It would be expected to decrease initially, for instance, for materials that exhibit the phenomenon of "work hardening." Certain solid state electronic devices are also believed to have a decreasing failure rate. There is some empirical evidence to support the assertion that in many instances the initial failure rate for a new component is nonzero (Davis, 1952).

A mathematical model sustained by some empirical evidence and considerable mathematical argument has been advanced to support the plausibility of the exponential distribution (see Section 3). Consider a large piece of equipment, such as an electronic computer. Suppose that equipment failure results from the failure of any component. If each component is replaced immediately upon failure and the lifetimes of components are stochastically independent, the sequence of equipment failures in time is simply the sequence of all individual component failures. Under fairly weak conditions on the component failure distributions, the sequence of equipment failures after a long time is approximately a Poisson process if the number of components is large (see Theorem 3.2). This result will hold if, for example, the average life of all components is uniformly bounded from below by some positive number and the component failure distributions have increasing failure rate. Because the time between successive events in a Poisson process is exponential and these times are independent, this result is often invoked to support the assumption of an exponential failure law.

It can be shown that this model is equivalent to the telephone trunking situation in which a large number of subscribers are feeding into a telephone exchange. The time between calls tends to an exponential distribution after a long time if the number of subscribers feeding into the exchange is large. A particularly convenient feature of this result is that the exponential parameter depends only on the mean of the individual component failure distributions. However, to answer questions concerning optimum individual component replacement policies, it is usually necessary to know more than the mean of the failure distribution. Except in isolated instances, the exponential does not appear as the distribution of first time to failure for equipment in which all components are initially new.

In general, except for large equipments with component renewal, a variety of families of distributions have been used in different applications. Typical failure distributions finding some favor are the gamma, the Weibull, the normal, and the log normal (see Section 2). All these families, excepting the log normal, have increasing failure rate for some parameter values and are therefore at least reasonable candidates. The failure rate of the log normal increases at first and then eventually decreases to zero. For this reason the log normal has found some disfavor as a failure distribution (Freudenthal, 1960). It has been proposed as a reasonable family of distributions for describing the length of time to repair a piece of equipment, however. There is some empirical evidence for this assertion (Green and Horne, 1961). We might also argue that if a repair has not been completed after a long time, there is less chance of its being immediately completed because of psychological and logistic factors. For example, a repairman may become discouraged after a long struggle, or excessive repair time may be due to nonavailability of a spare part.

So far we have been discussing the properties of selected families of failure distributions. Perhaps a more fruitful avenue of investigation is to study the properties of certain nonparametric classes based on physical considerations. Intuitively we expect the conditional mean residual life of a used component to be less than that of a new component. Assuming an even stronger condition, we say that a distribution F has *decreasing mean residual life* if and only if the mean residual life of a component of age t,

$$\frac{\int_t^\infty \bar{F}(x)\,dx}{\bar{F}(t)}, \tag{1.1}$$

is decreasing in t. Weak as this assumption appears, it is still possible to deduce some rather interesting properties enjoyed by this class of distributions (Barlow, Marshall, Proschan, 1963).

A natural strengthening of the concept of decreasing mean residual life is to assume that the conditional probability of failure given survival to time t is increasing in t; that is, $r(t)$ is increasing in t. Distributions with this property will be denoted IFR for *increasing failure rate*. Distributions for which $r(t)$ is decreasing will be denoted DFR for *decreasing failure rate*. Under either assumption we can obtain sharp bounds on the survival probability in terms of moments and percentiles. Upper and lower bounds on survival probability given the first two moments are tabulated in Appendix 3. These bounds are much tighter than the classical Chebyshev bounds without the IFR or DFR assumption. IFR and DFR distributions are also preserved under various types of structuring. Because of these properties and the intuitive appeal of IFR and DFR distributions in many applications, the IFR or DFR assumption will often be invoked in succeeding chapters. A statistical test is given in Appendix 2 for testing the hypothesis that F is exponential against the alternative that F is IFR (DFR) and not exponential.

2. TYPICAL FAILURE LAWS

Several functions are used and are equally suitable for describing the failure distribution, namely, the probability density $f(t)$ when it exists, the cumulative probability $F(t)$, and the failure rate $r(t)$.

These quantities are related by

$$F(t) = \int_0^t f(x)\, dx$$

$$\bar{F}(t) = \exp\left[-\int_0^t r(x)\, dx\right] \qquad (2.1)$$

where $\bar{F}(t) = 1 - F(t)$. It will be assumed throughout the book unless otherwise specified that

$$F(0^-) = 0, \quad F(+\infty) = 1$$

and that F is right continuous.

A variety of families of distributions have been used for the fatigue failure of materials and the life length of electronic and mechanical components. Typical useful failure laws which have been assumed are given next. The failure rate is given when it is a simple expression.

(i) The exponential:
$$f(t) = \lambda e^{-\lambda t}; \quad r(t) = \lambda, \quad \lambda > 0, t \geq 0.$$

(ii) The gamma:
$$f(t) = \lambda(\lambda t)^{\alpha-1} e^{-\lambda t}/\Gamma(\alpha), \quad \lambda, \alpha > 0, t \geq 0.$$

(iii) The Weibull:
$$f(t) = \lambda \alpha t^{\alpha-1} e^{-\lambda t^\alpha}, \quad r(t) = \lambda \alpha t^{\alpha-1}, \quad \lambda, \alpha > 0, t \geq 0.$$

(iv) The modified extreme value distribution:
$$f(t) = \frac{1}{\lambda} \exp\left[-\frac{e^t - 1}{\lambda} + t\right], \quad r(t) = \frac{e^t}{\lambda}, \quad \lambda > 0, t \geq 0.$$

(v) The truncated normal:
$$f(t) = \frac{1}{a\sigma\sqrt{2\pi}} \exp\left[-\frac{(t-\mu)^2}{2\sigma^2}\right],$$
$$\sigma > 0, -\infty < \mu < \infty, 0 \leq t < \infty,$$

where a is a normalizing constant.

(vi) The log normal:
$$f(t) = \frac{1}{t\sigma\sqrt{2\pi}} \exp\left[-\frac{1}{2\sigma^2}(\log t - \mu)^2\right],$$
$$-\infty < \mu < \infty, \sigma > 0, t \geq 0.$$

Generalizations of (i), (ii), (iii), and (iv) can be achieved by replacing t by $t - a$, where a represents the so-called "guarantee" time; that is, failure cannot occur before a units of time have elapsed.

The exponential distribution has constant failure rate, but the gamma (Figure 2.1) and Weibull (Figure 2.2) distributions have increasing failure rate for $\alpha > 1$. The modified extreme value distribution and the normal (Figure 2.3) have increasing failure rate. The log normal distribution has a decreasing failure rate in the long-life range. It appears doubtful, therefore, that the log normal distribution is physically relevant to such phenomena as fatigue in materials.

The family of exponential distributions is the best known and most thoroughly explored, largely through the work of B. Epstein and his associates (Epstein, 1958). The exponential distribution has a number of desirable mathematical properties, but its applicability is limited because of the following property: it can be proved that if the life length T of a structure has the exponential distribution, previous use does not affect its future life length. In other words, if a structure has not failed up to a

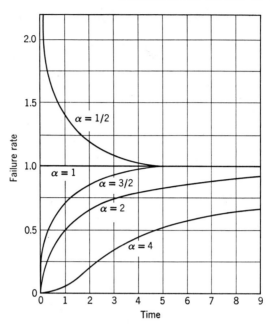

Fig. 2.1. Failure rate curves of the gamma distribution for $\lambda = 1$.

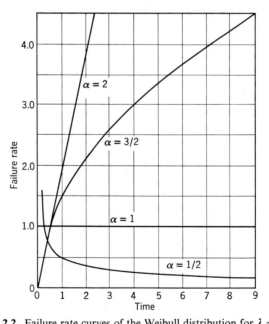

Fig. 2.2. Failure rate curves of the Weibull distribution for $\lambda = 1$.

FAILURE DISTRIBUTIONS 15

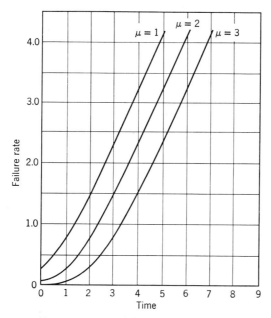

Fig. 2.3. Failure rate curves of the normal distribution for $\sigma = 1$.

time t, the probability distribution of its future life length $T - t$ is the same as if the structure were quite new and had just been placed in use at time t. The exponential distribution is the only distribution with this property (Feller, 1957, p. 413). There are structures which have this property, for example, an electric fuse (assuming it cannot melt partially) whose future life distribution is practically unchanged as long as it has not yet failed. As another example, imagine a situation in which a device under test is subject to an environment that can be described by some sort of random process. Let us imagine that this random process has peaks generated according to a Poisson process, and that it is only these peaks that can affect the device, in the sense that the device will fail if a peak occurs and will not fail otherwise. If the Poisson rate is λ, the failure distribution for the device will be given by (i) and in this situation previous use does not affect future life length. It is clear, however, that this property of the exponential distribution makes it inadequate for describing the life distribution of any structure which, when in normal use, undergoes changes affecting its future life length. A notable exception occurs in the case of a complex structure whose components are replaced. The time between structure failures may be approximately exponentially distributed (see Section 3).

Davis (1952) has examined failure data for a wide variety of items. Certain electronic components, such as the fuses mentioned earlier, seem to follow an exponential failure law. Davis notes, however, that for those items under close control of both the manufacturing process and the conditions of test, a *normal* theory of failure seems to be consistent with the data. However, many life length distributions occurring in practical applications are obviously not normal because they are markedly skewed whereas the normal distribution is symmetrical.

The *gamma* family of distributions is skewed and therefore may seem more natural than the normal family. The members of this family have increasing failure rate for $\alpha > 1$ and, in this case, the failure rate is bounded above by λ; for $\alpha < 1$, the failure rate is decreasing. Assume that a device subject to a given environment will fail when exactly $k \geq 1$ shocks occur and not before. If shocks occur at a Poisson rate λ, the waiting time (or life) until the item fails is described by

$$f(t) = \lambda^k t^{k-1} e^{-\lambda t}/\Gamma(k), \quad t \geq 0,$$

since this is the density of the convolution of k exponentials.

The *Weibull* family of distributions has increasing failure rate for $\alpha > 1$ and, in this case, the failure rate is unbounded. This family is called the Type III asymptotic distribution of extreme values by Gumbel (1958). Let $X_1, X_2, \ldots, X_n$ be independent observations on a random variable X with distribution F and let $Y_n = \min(X_1, \ldots, X_n)$. Suppose $F(x) \sim cx^\alpha$ as x tends to 0 ($\alpha > 0$). Then $(nc)^{1/\alpha} Y_n$ has a limiting Weibull distribution, that is,

$$P[(nc)^{1/\alpha} Y_n \geq t] \sim e^{-t^\alpha}$$

(Cox, 1962, p. 109). A more general form of its density is given by (iii). In (iii) the failure rate

$$r(t) = \lambda \alpha t^{\alpha - 1}$$

is a power of t. Weibull felt that this was a good approximation for the failure rate, and it motivated his interest in this family. The Weibull distribution has been used to describe fatigue failure (Weibull, 1939), vacuum tube failure (Kao, 1958), and ball-bearing failure (Lieblein and Zelen, 1956). It is perhaps the most popular parametric family of failure distributions at the present time.

The *extreme value* distributions arise in the following way. Let $X_1, X_2, \ldots, X_n, \ldots$ be a sequence of independent observations of a random variable X. Let

$$Y_1 = X_1, \quad Y_2 = \max(X_1, X_2), \ldots, Y_n = \max(X_1, X_2, \ldots, X_n), \ldots,$$

and let $Z_1 = X_1$, $Z_2 = \min(X_1, X_2), \ldots, Z_n = \min(X_1, X_2, \ldots, X_n)$.

FAILURE DISTRIBUTIONS 17

The random variables Y_n and Z_n are the extreme values (respectively, the maximum and the minimum) of the sample of independent observations $X_1, X_2, \ldots, X_n$. The asymptotic distribution (that is, the distribution for large values of n) of the extreme values Y_n and Z_n can be shown to be one of several types, depending on the behavior of the distribution function of the parent random variables $X_1, X_2, \ldots, X_n, \ldots$. A complete discussion of the asymptotic distribution of extreme values is given by E. J. Gumbel (1958). What we have called a modified extreme value distribution is a modification of the Type I asymptotic distribution of extreme values given by

$$F(t) = 1 - \exp\left(-e^{\lambda(t-\mu)}\right),$$

where $-\infty < \mu < \infty$, $\lambda > 0$, and $-\infty < t < \infty$.

As noted earlier, the *log normal* distribution seems to have little except its mathematical tractability to recommend itself as a failure distribution. The log normal does seem to give a good fit to repair time distributions, however, and this will be a subject of some interest in subsequent chapters.

In fatigue studies the time to failure is measured in the number of cycles to failure and is, therefore, a discrete random variable. Thus it will be useful to have some typical discrete failure distributions on record.

Typical discrete failure distributions are

(vii) the geometric:

$$p_k = p(1-p)^k, \quad 0 < p < 1, k = 0, 1, 2, \ldots.$$

(viii) The negative binomial (or Pascal):

$$p_k = \binom{-\alpha}{k} p^\alpha (-q)^k, \quad p = 1 - q > 0, \quad \alpha \geq 0,$$

$$k = 0, 1, 2, \ldots.$$

(ix) The binomial:

$$p_k = \binom{n}{k} p^k (1-p)^{n-k}, \quad n = \text{positive integer}, 0 < p < 1,$$

$$k = 0, 1, 2, \ldots, n.$$

(x) The Poisson:

$$p_k = \frac{\lambda^k \exp(-\lambda)}{k!}, \quad \lambda > 0, k = 0, 1, 2, \ldots.$$

There is, of course, a natural analogue to $r(t)$ for discrete distributions. The failure rate function of a discrete distribution $\{p_k\}_{k=0}^{\infty}$ is

$$r(k) = \frac{p_k}{\sum_{j=k}^{\infty} p_j}$$

Note that in this case $r(k) \leq 1$. The negative binomial distributions have increasing failure rate for $\alpha > 1$ and decreasing failure rate for $\alpha < 1$. This family coincides with the geometric family for $\alpha = 1$ and, in this case, the failure rate is constant. The binomial and Poisson distributions have increasing failure rate.

3. THE EXPONENTIAL AS THE FAILURE LAW OF COMPLEX EQUIPMENT

As noted in Section 2, equipment failing according to an exponential distribution essentially does not age. For this reason the exponential distribution was considered to be of limited applicability. There is, however, another situation in which the exponential distribution plays a prominent role. Consider a system consisting of many components, each subject to an individual pattern of malfunction and replacement and all parts making up the failure pattern of the equipment as a whole. Under some reasonably general conditions, the distribution of the time between equipment failures tends to the exponential as the complexity and the time of operation increase.

Imagine a complex piece of equipment as a large number of sockets into each of which there is inserted a component. After the equipment is put into operation, it remains in operation continuously. We make the following assumptions.

1. Components in different sockets are not necessarily alike and are stochastically independent.
2. Every component failure causes equipment failure.
3. Each component is replaced immediately at failure.
4. The process of failure detection, trouble location, and replacement is assumed to consume no appreciable time (or if it does, that time is taken to be part of the time between failures).

We are interested in the distribution of the time between failures after a long period of time has elapsed and as n (the number of components) becomes very large. The method of solution is first to let the operating

FAILURE DISTRIBUTIONS   19

time approach infinity (i.e., consider the stationary case) and then let n become infinite.

As time goes on, there develops at each socket an unending sequence of failures which constitutes a random process called a "renewal process" (see Chapter 3). The overall failure pattern of the equipment is therefore a superposition of many such renewal processes. Associated with the renewal process corresponding to the ith component is a random variable $N_{ni}(t)$, denoting the number of failures of the ith component in time t for a structure of n components. If the equipment has been put in operation at time $-\infty$, the number of failures in $[0, t]$ depends only on t and this is considered to be the stationary situation.

In Chapter 3, Section 2, it is shown that for the stationary sequence of failures in the ith component,

$$P[N_{ni}(t) > 0] = \int_0^t \frac{\bar{F}_{ni}(x)}{m_{ni}} \, dx$$

where $F_{ni}(t)$ is the failure distribution of the ith component and m_{ni} is its mean in a structure of size n. The parameter of the ith component sequence is defined as the following limit:

$$\lim_{t \to 0} \frac{P[N_{ni}(t) > 0]}{t} = \frac{1}{m_{ni}}.$$

We always assume that the probability of two or more failures occurring at the same time is zero. The condition on the component processes in the stationary situation to ensure that the sum process approaches a Poisson process can be stated in terms of $F_{ni}(t)$.

Now we can formulate our problem exactly. Let

$N_{11}(t)$;
$N_{21}(t), N_{22}(t)$;
. . . .
$N_{n1}(t), N_{n2}(t), \ldots, N_{nn}(t)$
. . . .

be a double sequence of processes which have been going on since time $-\infty$, where the processes are independent within each row. By $N_n(t)$ we shall denote their superposition (a simple sum) in the nth row. Adding the parameters of the processes in a given row, we obtain the parameter

$$\frac{1}{\Lambda_n} = \frac{1}{m_{n1}} + \frac{1}{m_{n2}} + \cdots + \frac{1}{m_{nn}}$$

of the sum process. For simplicity we shall suppose $\Lambda_n = \Lambda$ remains constant as n increases.

The problem consists in finding restrictions on the distributions F_{ni} such that with increasing n the sum process $[N_n(t); t \geq 0]$ converges to the Poisson process with parameter $1/\Lambda$. The following theorem is from Ososkov (1956). We shall not present the proof here. For a more complete discussion of a slightly different version of this theorem, see A. Ya. Khintchine (1960, Chap. V).

THEOREM 3.1 (Ososkov). If

(i) $\Lambda \sum_{i=1}^{n} \dfrac{F_{ni}\left(\frac{t}{\Lambda}\right)}{m_{ni}} \to 1, \quad n \to \infty, t > 0$ (3.1)

(ii) $[N_{ni}(t); t \geq 0]$ is a stationary renewal process for $i = 1, 2, \ldots, n$;

(iii) $N_n(t) = \sum_{i=1}^{n} N_{ni}(t)$,

then

$$\lim_{n \to \infty} P[N_n(t_i) = k_i, 1 \leq i \leq s] = e^{-\lambda t_s} \lambda^{k_s} \prod_{i=1}^{s} \dfrac{(t_i - t_{i-1})^{k_i - k_{i-1}}}{(k_i - k_{i-1})!} \quad (3.2)$$

where $\lambda = 1/\Lambda$, $t_0 = k_0 = 0$, s is any positive integer, $0 < t_1 < t_2 < \cdots < t_s$; $0 \leq k_1 \leq k_2 \leq \cdots \leq k_s$; and all k_i are nonnegative integers (that is, $[N_n(t); t \geq 0]$ converges to a Poisson process with parameter $1/\Lambda$).

If (ii), (iii), and (3.2) hold and if for arbitrarily small $\sigma > 0$ there exists n_0 such that

$$\max_{1 \leq i \leq n} \dfrac{1}{m_{ni}} < \sigma \quad (3.3)$$

for all $n \geq n_0$, then (3.1) follows.

It can be shown that the infinitesimal smallness of the parameters of the added processes is an immediate consequence of (3.1). Thus requirement (3.3) enters into the necessity as well as into the sufficiency conditions of the theorem; however, in the latter case (3.3) enters implictly because not (3.1) but a consequence of (3.1) is used. This requirement of infinitesimal smallness of the parameters of the added processes is quite reasonable. It removes the trivial possibility that any process $N_{ni}(t)$ with parameter $1/\Lambda$ could be the limit for $N_n(t)$ (for this it is sufficient, for example, to take in each sequence the process $N_{ni}(t)$ itself as the first summand and to take for the remaining summands fictitious processes which contribute no failures). The role of condition (3.1) is to prevent the possibility that a large number of failures of the same process accumulates in small regions (this can give multiple jumps in the limit process, thus preventing the convergence of the sum process to a Poisson process).

FAILURE DISTRIBUTIONS

Note that instead of supposing that Λ_n remains constant, we could have normalized the summand process by dividing the mean of the component failure distribution by Λ_n. Then we would need only one index on our component failure distribution and

$$F_{ni}(t) = F_i(t\Lambda_n).$$

Without using double subscripts, we shall now state the theorem with stronger but more intuitive assumptions. First, note that by Chapter 3, Section 2, in the stationary situation the probability of no equipment failures in $[0, t]$ is

$$\bar{G}_n(t) = \prod_{i=1}^{n} \left[\int_t^\infty \frac{\bar{F}_i(x)\,dx}{m_i} \right] \tag{3.4}$$

for an equipment with n components. The proof of the following theorem is provided by R. Drenick (1960b).

THEOREM 3.2 (Drenick). If

(i) $\lim\limits_{n\to\infty} \sup\limits_{1\leq i\leq n} \dfrac{\Lambda_n}{m_i} = 0$;

(ii) $F_i(t) \leq At^\sigma$, $A > 0$, $\sigma > 0$, as $t \to 0$, $i = 1, 2, \ldots, n$;

then

$$\lim_{n\to\infty} \bar{G}_n(\tau\Lambda_n) = e^{-\tau} \quad \text{for} \quad \tau > 0,$$

where $\bar{G}_n(\tau)$ is the *stationary* probability that an equipment with n components survives the interval $[0, \tau]$.

These conditions, though more intuitive than those in Theorem 3.1, are, of course, more than sufficient. Condition (i) requires that $\Lambda_n \to 0$ as $n \to \infty$. It also requires, roughly speaking, that there not be any very "poor" components in the equipment, namely, components with mean lives so small that Λ_n/m_i increases with n even though Λ_n decreases. This would certainly be so if, for instance, the m_i were bounded below. Condition (ii), from Cox and Smith (1954), controls the behavior of the failure law near $t = 0$. It excludes failure rates which go to infinity, as $t \to 0$, at least as rapidly as $1/t$, but includes all failure laws that are commonly used. If, for example, $1/m_i \leq A$ and F_i has increasing failure rate for all i, then $F_i(t) \cong f_i(0)\,t \leq t/m_i \leq At$ [see Equation (7.1)] in a neighborhood of the origin and all conditions of Theorem 3.2 are satisfied.

We can ask whether a similar result can be obtained for the initial survival probability (i.e., nonstationary case). This is the probability that the equipment will operate without a failure for a period $[0, t]$ after it has first been put into operation. In general, much more stringent conditions must be imposed on the component failure distributions to assure convergence of the equipment distribution as $n \to \infty$. The Weibull class of distributions, which, of course, contains the exponential, seems to be the smallest class of interest here. If the component failure distributions behave like βt^α for some α, $\beta > 0$ in a neighborhood of the origin, then

$$\bar{G}_n(t) \cong e^{-n\beta t^\alpha}$$

for large n. For a detailed discussion of extremal distributions, see Gumbel (1958).

4. MONOTONE FAILURE RATES

Since most materials, structures, and devices wear out with time, the class of failure distributions for which $r(t)$ is increasing is one of special interest. The phenomenon of work hardening of certain materials and the "debugging" of complex systems make the class of failure distributions with decreasing failure rate also of some interest. Here and in the following, the terms "increasing" and "decreasing" are *not* used in the strict sense. Note that with this convention the exponential distribution with constant failure rate belongs to both classes. There are, of course, examples, such as the dynamic loading of structures, where a nonmonotonic failure rate function would be appropriate. Structures undergoing adjustment and modification would also tend to have a nonmonotonic failure rate.

Distributions with monotone failure rate can arise in other contexts than that of time to failure. For example, suppose a structure is subjected to a constantly increasing stress so that the stress level at which failure occurs will in general be a random variable, say X. Because the conditional probability of failure at stress level t given survival to stress level $s < t$ should intuitively increase with s, the distribution of X will be IFR.

The assumption that a failure distribution has a monotone failure rate function is quite strong, as we shall show. Such failure distributions possess many useful and interesting qualitative properties. The results presented here will be exploited in subsequent chapters. The material for this section is based on Barlow, Marshall, and Proschan (1963) and Barlow and Marshall (1964).

Definitions

Failure rate is sometimes defined as the probability of failure in a finite interval of time, say of length x, given the age of the component, say t.

FAILURE DISTRIBUTIONS

If F denotes the failure distribution, then the failure rate, by this definition would be
$$\frac{F(t+x) - F(t)}{\bar{F}(t)}. \tag{4.1}$$

If we divide this quantity by x and let $x \to 0$, we obtain $r(t)$. In many contexts (Herd, 1955) $r(t)$ is called the "hazard rate" or "instantaneous failure rate." In this monograph we do not make this distinction. It will be convenient to define IFR (DFR) distributions in terms of the ratio (4.1) rather than $r(t)$, however. We must first distinguish between discrete and nondiscrete distributions. A distribution is discrete if it concentrates all probability on, at most, a countable set of points.

DEFINITION 1. A nondiscrete distribution F is IFR (DFR) if and only if
$$\frac{F(t+x) - F(t)}{\bar{F}(t)} \tag{4.1}$$
is increasing (decreasing) in t for $x > 0$, $t \geq 0$ such that $F(t) < 1$.

We could have defined IFR without restricting t to nonnegative values; however, for DFR, we cannot extend t to $-\infty$. Note that if F is DFR, then $F(t) > 0$ for $t > 0$. The convenience of Definition 1 is that we need not assume that F has a density. A distribution can be IFR (DFR) and have a jump at the right- (left-) hand end point of its interval of support. The following theorem shows for distributions on the positive half of the axis the equivalence of IFR (DFR) distributions with a density and distributions for which $r(t)$ is increasing (decreasing).

LEMMA 4.1. Assume F has a density f, with $F(0^-) = 0$. Then F is IFR (DFR) if and only if $r(t)$ is increasing (decreasing).

Proof. Note that if we divide (4.1) by x and let x approach zero, we obtain $r(t)$. Hence we need only show that $r(t)$ increasing (decreasing) in t implies (4.1) increasing (decreasing) in t. For $x_1 \leq x_2$
$$r(x_1) \underset{(\geq)}{\leq} r(x_2)$$
implies
$$\int_0^t r(x_1 + u)\,du \underset{(\geq)}{\leq} \int_0^t r(x_2 + u)\,du.$$
That is,
$$\exp\left[-\int_{x_2}^{x_2+t} r(u)\,du\right] \underset{(\geq)}{\leq} \exp\left[-\int_{x_1}^{x_1+t} r(u)\,du\right]$$
which implies
$$\frac{F(x_2 + t) - F(x_2)}{\bar{F}(x_2)} \underset{(\leq)}{\geq} \frac{F(x_1 + t) - F(x_1)}{\bar{F}(x_1)}$$

using the identity

$$\bar{F}(t) = \exp\left[-\int_0^t r(u)\, du\right]. \; \|$$ (4.2)

Definition 1 has a discrete analogue.

DEFINITION 2. A discrete distribution $\{p_k\}_{k=0}^\infty$ is IFR (DFR) if and only if

$$\frac{p_k}{\sum_{i=k}^\infty p_i}$$

is nondecreasing (nonincreasing) in $k = 0, 1, 2, \ldots$.

A natural generalization of the ratio $f(t)/\bar{F}(t)$ is

$$\frac{f(t)}{F(t+\Delta) - F(t)}$$ (4.3)

Densities f for which (4.3) is nondecreasing in t wherever the denominator is nonzero for all real Δ are called Pólya frequency densities of order 2. This assumption will be imposed on the failure density in models which will be treated later. The definition of Pólya frequency function was originally stated in terms of determinants, which is in some applications more useful than the assertion that (4.3) is increasing.

DEFINITION 3. A function $p(x)$ defined for x in $(-\infty, \infty)$ is a Pólya frequency function of order 2 (PF_2) if and only if $p(x) \geq 0$ for all x and

$$\begin{vmatrix} p(x_1 - y_1) & p(x_1 - y_2) \\ p(x_2 - y_1) & p(x_2 - y_2) \end{vmatrix} \geq 0$$

whenever $-\infty < x_1 \leq x_2 < \infty$ and $-\infty < y_1 \leq y_2 < \infty$. (See Appendix 1 for a discussion of Pólya frequency functions, their properties, and their applications.) It is easy to check that $p(x)$ is PF_2 if and only if $p(x) \geq 0$ for all x and

$$\frac{p(x - \Delta)}{p(x)}$$ (4.4)

is increasing in x for $\{x \mid p(x) > 0\}$ and $\Delta > 0$. Using this result we show in Theorem 1 of Appendix 1 that f is a PF_2 density if and only if (4.3) is increasing in t.

FAILURE DISTRIBUTIONS

The concept of a PF_2 function has a natural extension, namely:

DEFINITION 4. A function $p(x, y)$ defined for $x \in X$ and $y \in Y$ (X and Y linearly ordered sets) is *totally positive* of order 2 (TP_2) if and only if $p(x, y) \geq 0$ for all $x \in X$, $y \in Y$ and

$$\begin{vmatrix} p(x_1, y_1) & p(x_1, y_2) \\ p(x_2, y_1) & p(x_2, y_2) \end{vmatrix} \geq 0$$

whenever $x_1 \leq x_2$ and $y_1 \leq y_2$ ($x_1, x_2 \in X$; $y_1, y_2 \in Y$). This concept was first defined and studied by S. Karlin and H. Rubin (1956). Further implications of TP_2 functions are discussed in Appendix 1. PF_2 functions are TP_2 in translation. We shall show that F is IFR (DFR) if and only if $\bar{F}(x - y)$ ($\bar{F}(x + y)$) is TP_2 in real x, y ($x + y \geq 0$). Each of the following equivalent definitions of distributions with monotone hazard rate will be important later.

THEOREM 4.1. The following statements are equivalent. We assume $F(0^-) = 0$.
(a) F is an IFR (DFR) distribution.
(b) Log $\bar{F}(t)$ is concave (convex) for t in $\{t \mid F(t) < 1, t \geq 0\}$.
(c) $\bar{F}(t)$ is PF_2 ($\bar{F}(x + y)$ is TP_2 in x and y for $x + y \geq 0$).
Proof. (a) $\Leftrightarrow$ (b)
Let $\bar{F}(t) = e^{-R(t)}$. Then

$$\frac{F(t + \Delta) - F(t)}{\bar{F}(t)} = 1 - e^{-[R(t+\Delta)-R(t)]}$$

and F is IFR (DFR) if and only if $R(t + \Delta) - R(t)$ is increasing (decreasing) in t for all $\Delta > 0$. Thus F is IFR (DFR) if and only if $R(t)$ is convex (concave).

(a) $\Leftrightarrow$ (c) F is IFR (DFR) if and only if

$$\begin{vmatrix} \bar{F}(t_1) - \bar{F}(t_1 + x) & \bar{F}(t_2) - \bar{F}(t_2 + x) \\ \bar{F}(t_1) & \bar{F}(t_2) \end{vmatrix} \underset{(\geq)}{\leq} 0$$

for $t_1 \leq t_2$ and $x \geq 0$. Subtracting the second row from the first, we see that this is the condition that $\bar{F}$ be PF_2 (TP_2 in sums of t for $t \geq 0$). ∥

A theorem analogous to Theorem 4.1 can be proved for discrete distributions which are IFR (DFR).

Since a measurable convex function is continuous in the interior of the region of its definition, we see from (b) of Theorem 4.1 that F cannot have a jump in the interior of its support if F is IFR or DFR. If F is IFR, F may possess a jump only at the right-hand end of its interval of support and therefore $F(0^-) = F(0) = 0$. If F is DFR, F may possess a jump only at the origin. If F is DFR with density f, then f is necessarily decreasing, because if f were to increase, so would $f(t)/\bar{F}(t)$. In addition, the support of f in this case must be the half-open interval $[0, \infty)$. For if the support of f were a finite interval, say $[0, b]$, then [using (4.2)]

$$\lim_{t \to b} \left\{ 1 - \exp \left[-\int_t^{t+\Delta} r(x)\, dx \right] \right\} = \lim_{t \to b} \frac{F(t + \Delta) - F(t)}{\bar{F}(t)} = 1$$

and hence $\overline{\lim}_{t \to b} r(t) = \infty$. But this contradicts the assumption that r is decreasing.

For F IFR (DFR) we have seen that F is continuous except possibly at the right- (left-) hand end point of its interval of support. Actually, we may show that in either case the continuous part of F is absolutely continuous. Suppose F is IFR. Let $\epsilon > 0$; choose z such that $U(z) = -\log \bar{F}(z) < \infty$. Let $0 \leq \alpha_1 < \beta_1 < \alpha_2 < \beta_2 < \cdots < \alpha_m < \beta_m \leq z$ be points satisfying $\sum_{i=1}^{m}(\beta_i - \alpha_i) < \epsilon/U^+(z)$, where

$$U^+(z) = \lim_{\delta \downarrow 0} [U(z + \delta) - U(z)]/\delta < \infty$$

because U is convex. Then

$$\sum_{i=1}^{m} |U(\beta_i) - U(\alpha_i)|$$
$$= \sum_{i=1}^{m} \frac{U(\beta_i) - U(\alpha_i)}{\beta_i - \alpha_i} (\beta_i - \alpha_i) \leq U^+(z) \sum_{i=1}^{m} (\beta_i - \alpha_i) \leq \epsilon.$$

Thus U is absolutely continuous on $[0, z]$, and the result follows. A similar argument applies if F is DFR.

Comparison with the exponential distribution

Because the exponential distribution with constant failure rate is the boundary distribution between IFR and DFR distributions, it provides natural bounds on the survival probability of IFR and DFR distributions. In particular, if F is IFR, $\bar{F}(t)$ crosses the function $e^{-\alpha t}$ at most once for $\alpha > 0$ unless they coincide identically. This is easily seen because the two functions agree at the origin and

$$\log \bar{F}(t) - (-\alpha t)$$

can vanish at most once for $t > 0$. Let F have mean μ_1; then $\bar{F}(t)$ must cross e^{-t/μ_1} exactly once, and the crossing is necessarily from above. This observation tells us that $\bar{F}(t)$ tails off exponentially fast. The following lemma is a more precise statement of this exponential rate of decrease.

FAILURE DISTRIBUTIONS 27

LEMMA 4.2. If F is IFR (DFR), then

$$[\bar{F}(t)]^{1/t}$$

is decreasing (increasing) in t.

Proof. Suppose F is IFR (DFR). Then $\log \bar{F}(t)$ is concave (convex) in t and

$$\frac{\log \bar{F}(t) - \log \bar{F}(0^-)}{t - 0}$$

is decreasing (increasing) in t and consequently $[\bar{F}(t)]^{1/t}$ is also decreasing (increasing) in t. ||

By Lemma 4.2, $\bar{F}(x) \leq [\bar{F}(t)]^{x/t}$ for $x > t$ and

$$\int_t^\infty x^r \bar{F}(x)\, dx \leq \int_t^\infty x^r [\bar{F}(t)]^{x/t}\, dx < \infty$$

when $\bar{F}(t) < 1$ and $r \geq 0$. Hence, IFR distributions have finite moments of all orders.

THEOREM 4.3. If F is IFR (DFR) and $F(\xi_p) = p$, that is, ξ_p is a pth percentile, then

$$\bar{F}(t) \begin{array}{c} \geq \\ (\leq) \end{array} e^{-\alpha t}, \quad t \leq \xi_p$$

$$\bar{F}(t) \begin{array}{c} \leq \\ (\geq) \end{array} e^{-\alpha t}, \quad t \geq \xi_p \qquad (4.5)$$

where

$$\alpha = -\frac{\log(1-p)}{\xi_p}$$

Proof. The inequalities are an immediate consequence of Lemma 4.2. ||

A remarkable lower bound on $\bar{F}(t)$ is true when F is IFR, namely

THEOREM 4.4. If F is IFR with mean μ_1, then

$$\bar{F}(t) \geq \begin{cases} e^{-t/\mu_1}, & t < \mu_1 \\ 0, & t \geq \mu_1 \end{cases} \qquad (4.6)$$

The inequality is sharp.

Proof. Because we can always approximate an IFR distribution arbitrarily closely by a continuous IFR distribution, we may assume without loss of generality that F is continuous. Let X be a random variable with distribution F. Then $\log \bar{F}(t)$ is concave in t and by Jensen's inequality

$$E[\log \bar{F}(X)] \leq \log \bar{F}(\mu_1).$$

Because F is continuous, $F(X)$ is a uniform random variable on $[0, 1]$ and

$$E[\log \bar{F}(X)] = \int_0^1 \log u\, du = -1.$$

Hence

$$\log \bar{F}(\mu_1) \geq -1$$

or

$$\bar{F}(\mu_1) \geq e^{-1}.$$

By Lemma 4.2

$$[\bar{F}(t)]^{1/t} \geq [\bar{F}(\mu_1)]^{1/\mu_1}$$

for $t < \mu_1$ and, therefore,

$$\bar{F}(t) \geq e^{-t/\mu_1}$$

for $t < \mu_1$.

The exponential distribution with mean μ_1 attains the lower bound for $t < \mu_1$. The degenerate distribution concentrating at μ_1 attains the lower bound for $t \geq \mu_1$. ∥

Note that the inequality is actually strict for $0 < t < \mu_1$, unless F coincides identically with e^{-t/μ_1}.

These bounds have obvious applications. For example, suppose a system consists of n independent components in series with IFR distributions F_i with corresponding means μ_i $(i = 1, 2, \ldots, n)$. Then by using (4.6) the system survival probability is

$$\prod_{i=1}^n \bar{F}_i(t) \geq \begin{cases} \exp\left(-t \sum_{i=1}^n \frac{1}{\mu_i}\right), & t < \min(\mu_1, \ldots, \mu_n) \\ 0, & \text{elsewhere.} \end{cases}$$

The bound is sharp, indicating why system reliability is often better than predicted on a parts count basis. Similarly, for a parallel system with IFR component distributions F_i with means μ_i $(i = 1, 2, \ldots, n)$

$$\prod_{i=1}^n F_i(t) \leq \begin{cases} \prod_{i=1}^n (1 - e^{-t/\mu_i}), & t < \min(\mu_1, \ldots, \mu_n) \\ 1, & \text{elsewhere.} \end{cases}$$

A generalization is obtained in Theorem 3.1 of Chapter 7. The best upper bound on $\bar{F}(t)$ when F is IFR is given by the following theorem.

THEOREM 4.5. If F is IFR with mean μ_1, then

$$\bar{F}(t) \leq \begin{cases} 1, & t \leq \mu_1 \\ e^{-\omega t}, & t > \mu_1 \end{cases} \quad (4.7)$$

where ω depends on t and satisfies $1 - \omega\mu_1 = e^{-\omega t}$.

FAILURE DISTRIBUTIONS

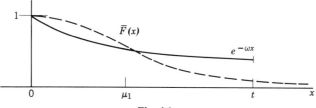

Fig. 4.1.

Proof. Note that $\bar{F}(x)$ can cross

$$\bar{G}(x) = \begin{cases} e^{-\omega x}, & x < t \\ 0, & x \geq t \end{cases}$$

at most once for $x < t$, and if it does cross, it crosses from above (Figure 4.1). If $t > \mu_1$, we can always determine ω so that $\int_0^t e^{-\omega x}\, dx = \mu_1$. Assume $G(x) \not\equiv F(x)$ for this choice of ω. Then $\bar{F}(x)$ cannot dominate $\bar{G}(x)$ for $0 \leq x \leq t$ because this would imply $\mu_1 = \int_0^\infty \bar{F}(x)\, dx > \int_0^t \bar{G}(x)\, dx$ $= \mu_1$. Therefore $\bar{F}(t) < e^{-\omega t}$ unless F coincides identically with G. Because G is IFR, the bound is sharp. The degenerate distribution concentrating at μ_1 provides the upper bound for $t \leq \mu_1$. ‖

As might be anticipated, the upper bound in Theorem 4.5 can be shown to be asymptotically equal to e^{-t/μ_1}. Figure 4.2 illustrates the best upper

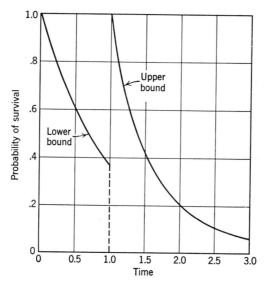

Fig. 4.2. Bounds on survival probability for IFR distributions ($\mu_1 = 1$).

and lower bounds on $\bar{F}(t)$ when F is IFR and $\mu_1 = 1$. Table I in Appendix 3 tabulates the upper bound. See Section 6 for generalizations of Theorems 4.4 and 4.5.

We can also obtain bounds on percentiles in terms of the mean and vice versa.

THEOREM 4.6. Assume F is IFR. If $p \leq 1 - e^{-1}$, then

$$[-\log(1-p)]\mu_1 \leq \xi_p \leq \left[-\frac{\log(1-p)}{p}\right]\mu_1;$$

if $p \geq 1 - e^{-1}$, then

$$\mu_1 \leq \xi_p \leq \left[-\frac{\log(1-p)}{p}\right]\mu_1,$$

where $\qquad \xi_p = \sup\,[t \mid F(t) \leq p]$.

The inequalities are sharp.

Proof. First we obtain the upper bound. By using Lemma 4.2,

$$\mu_1 = \int_0^\infty \bar{F}(x)\,dx \geq \int_0^{\xi_p} [\bar{F}(\xi_p)]^{x/\xi_p}\,dx \geq -\frac{\xi_p p}{\log(1-p)}.$$

Solving, we have

$$\xi_p \leq \mu_1 \frac{-\log(1-p)}{p}$$

for all $1 > p > 0$.

Assume $p \leq 1 - e^{-1}$. If $\xi_p < \mu_1$, then by Theorem 4.4, $1 - p = \bar{F}(\xi_p) \geq e^{-\xi_p/\mu_1}$, which implies $\xi_p \geq \mu_1[-\log(1-p)]$. Similarly, if $\xi_p \geq \mu_1$, $\bar{F}(\xi_p) \geq 1 - p \geq e^{-1} \geq e^{-\xi_p/\mu_1}$. Solving for ξ_p, we have

$$\xi_p \geq \mu_1[-\log(1-p)]$$

where $p \leq 1 - e^{-1}$.

Assume $p \geq 1 - e^{-1}$ and $\xi_p < \mu_1$. Then $F(\xi_p) = p$, because F cannot have a jump to the left of μ_1. Hence,

$$\bar{F}(\mu_1) \geq e^{-1} \geq 1 - p = \bar{F}(\xi_p)$$

implies $\xi_p \geq \mu_1$. From this contradiction it follows that if $p \geq 1 - e^{-1}$, then $\xi_p \geq \mu_1$.

IFR distributions with the prescribed percentiles attaining the bounds can be easily constructed from this argument. ∥

In life test samples, some items may not fail at all during the course of the test. Therefore, the usual sample average cannot be used. Some percentile estimates will always be available, however. Note that by using Theorem 4.6 we can obtain bounds on the mean of IFR distributions in

FAILURE DISTRIBUTIONS   31

terms of percentiles. For example, if M is the median, then

$$\frac{M}{2\log 2} \leq \mu_1 \leq \frac{M}{\log 2}.$$

For DFR distribution F, upper bounds on $\bar{F}(t)$ can be given in terms of a single moment, as shown in

THEOREM 4.7. If F is DFR with mean μ_1, then

$$\bar{F}(t) \leq \begin{cases} e^{-t/\mu_1}, & t \leq \mu_1 \\ \dfrac{\mu_1 e^{-1}}{t}, & t \geq \mu_1. \end{cases} \tag{4.8}$$

The inequality is sharp.

Proof. Let $L = \log \bar{F}(t)$. Because $\log \bar{F}(x)$ is convex, there exists α, $L \leq \alpha \leq 0$, such that

$$\log \bar{F}(x) \geq \frac{L-\alpha}{t} x + \alpha, \quad x \geq 0,$$

or

$$\bar{F}(x) \geq \exp\left(\frac{L-\alpha}{t} x + \alpha\right).$$

Thus for some α, $L \leq \alpha \leq 0$,

$$\mu_1 = \int_0^\infty \bar{F}(x)\, dx \geq \int_0^\infty \exp\left(\frac{L-\alpha}{t} x + \alpha\right) dx$$

$$= \frac{te^\alpha}{\alpha - L}$$

or

$$L \leq \alpha - \frac{te^\alpha}{\mu_1} = \varphi(\alpha).$$

Subject to $\alpha \leq 0$, $\varphi(\alpha)$ is maximized by

$$\alpha_0 = \begin{cases} -\log(t/\mu_1), & t \geq \mu_1 \\ 0, & t \leq \mu_1 \end{cases}$$

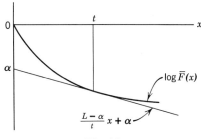

Fig. 4.3

and $L \leq \varphi(x_0)$ yields (4.8). Equality is attained uniquely by the distribution G, where

$$\log \bar{G}(x) = \begin{cases} 0, & x < 0 \\ -\dfrac{x}{t} + \log \dfrac{\mu_1}{t}, & x \geq 0 \end{cases} \quad \text{for } t \geq \mu_1. \quad \|$$

Bounds in terms of the rth moment can be easily obtained by the same method. Theorem 4.4 could also have been proved by the method of Theorem 4.7.

The sharp lower bound on $\bar{F}(t)$ for DFR distributions is zero. To see this, let $\epsilon > 0$ be arbitrarily small and

$$\bar{G}(x) = \begin{cases} 1, & x < 0 \\ \epsilon e^{-\alpha x}, & x \geq 0 \end{cases}$$

where $\alpha = \epsilon/\mu_1$. Then G is DFR with a jump at the origin and $\bar{G}(x) \leq \epsilon$ for all $x \geq 0$.

The following theorem will provide us with useful moment inequalities. For convenience we denote an exponential distribution by $G(t)$.

THEOREM 4.8. *If*
(a) F *is IFR with mean* μ_1 *and* $\bar{G}(x) = e^{-x/\mu_1}$,
(b) $\varphi(x)$ *is increasing (decreasing),*

then
$$\int_0^\infty \varphi(x) \bar{F}(x) \, dx \underset{(\geq)}{\leq} \int_0^\infty \varphi(x) \bar{G}(x) \, dx. \tag{4.9}$$

Proof. Suppose φ is increasing, and F is not identically equal to G. Because F is IFR and G is the exponential distribution with the same mean, $\bar{F}$ crosses $\bar{G}$ exactly once from above at, say, t_0; that is, $\bar{F}(t_0) = \bar{G}(t_0)$. Then

$$\int_0^\infty \varphi(x) \bar{F}(x) \, dx - \int_0^\infty \varphi(x) \bar{G}(x) \, dx$$
$$= \int_0^\infty [\varphi(x) - \varphi(t_0)][\bar{F}(x) - \bar{G}(x)] \, dx \leq 0$$

which proves (4.9). To prove the theorem for φ decreasing, replace φ by $-\varphi$. $\|$

Note that a similar theorem is true for DFR distributions with all inequalities reversed. From Theorem 4.8 we have an immediate comparison between the moments of an IFR distribution and the corresponding moments of an exponential distribution with the same mean.

FAILURE DISTRIBUTIONS

COROLLARY 4.9. If F is IFR (DFR) with rth moment μ_r, then

(a) $$\mu_r \begin{array}{c} \leq \\ (\geq) \end{array} \Gamma(r+1)\mu_1^r, \quad r \geq 1$$
$$\mu_r \begin{array}{c} \geq \\ (\leq) \end{array} \Gamma(r+1)\mu_1^r, \quad 0 \leq r \leq 1 \qquad (4.10)$$

(b) $$\int_0^\infty e^{-sx}\,dF(x) \begin{array}{c} \leq \\ (\geq) \end{array} \frac{1}{1+s\mu_1}. \qquad (4.11)$$

Proof. (a) Choose $\varphi(x) = x^{r-1}$ in Theorem 4.8. Then

$$\mu_r = r\int_0^\infty x^{r-1} F(x)\,dx \leq r\int_0^\infty x^{r-1} \bar{G}(x)\,dx = \Gamma(r+1)\mu_1^r,$$

for $r \geq 1$. Note that for $0 \leq r \leq 1$, $\varphi(x) = x^{r-1}$ is decreasing, so that the inequality is reversed.

(b) Because e^{-sx} is decreasing in x for $s > 0$ and

$$\int_0^\infty \frac{1-e^{-sx}}{s}\,dF(x) = \int_0^\infty e^{-sx} \bar{F}(x)\,dx \begin{array}{c} \geq \\ (\leq) \end{array} \int_0^\infty e^{-sx} \bar{G}(x)\,dx = \frac{1-1/(1+s\mu_1)}{s}$$

we have (4.11). ‖

Note that for an IFR distribution

$$\mu_2 \leq 2\mu_1^2$$

and, therefore, the variance σ^2 satisfies $\sigma^2 \leq \mu_1^2$ and the coefficient of variation σ/μ_1 is less than or equal to one. The inequalities are reversed for DFR distributions. From the proof of Corollary 4.9 we see that all inequalities are strict unless F coincides identically with the exponential distribution.

We now show that the mean life of a series system with IFR components whose means are μ_i ($i = 1, 2, \ldots, n$) exceeds the mean life of a series system with exponential components and means μ_i ($i = 1, 2, \ldots, n$). Just the reverse is true for a parallel system, however.

COROLLARY 4.10. If $F_i(x)$ is IFR (DFR) with mean μ_i and $\bar{G}_i(x) = e^{-x/\mu_i}$ ($i = 1, 2, \ldots, n$), then

(a) $$\int_0^\infty \prod_{i=1}^n F_i(x)\,dx \begin{array}{c} \geq \\ (\leq) \end{array} \int_0^\infty \prod_{i=1}^n \bar{G}_i(x)\,dx = \frac{1}{\sum_1^n 1/\mu_i}$$

(b) $$\int_0^\infty \left[1 - \prod_{i=1}^n F_i(x)\right] dx \begin{array}{c} \leq \\ (\geq) \end{array} \int_0^\infty \left[1 - \prod_{i=1}^n G_i(x)\right] dx.$$

Proof. First we show (*a*). By Theorem 4.8

$$\int_0^\infty \left[\prod_{j=1}^{i-1} \bar{F}_j(x) \prod_{j=i+1}^{n} \bar{G}_j(x)\right] F_i(x)\, dx \underset{(\leq)}{\geq} \int_0^\infty \left[\prod_{j=1}^{i-1} \bar{F}_j(x) \prod_{j=i+1}^{n} \bar{G}_j(x)\right] G_i(x)\, dx$$

for $1 \leq i \leq n$. By recursion we obtain

$$\int_0^\infty \prod_{j=1}^{n} \bar{F}_j(x)\, dx \underset{(\leq)}{\geq} \int_0^\infty \prod_{j=1}^{n} \bar{G}_j(x)\, dx = \frac{1}{\sum_{i=1}^{n} 1/\mu_i}.$$

To show (*b*), note that again by Theorem 4.8

$$\int_0^\infty \prod_{j=1}^{i-1} F_j(x) \prod_{j=i+1}^{n} G_j(x) \bar{F}_i(x)\, dx \underset{(\geq)}{\leq} \int_0^\infty \prod_{j=1}^{i-1} F_j(x) \prod_{j=i+1}^{n} G_j(x) \bar{G}_i(x)\, dx$$

which in turn implies

$$\int_0^\infty \left[1 - \prod_{j=1}^{i} F_j(x) \prod_{j=i+1}^{n} G_j(x)\right] dx \underset{(\geq)}{\leq} \int_0^\infty \left[1 - \prod_{j=1}^{i-1} F_j(x) \prod_{j=i}^{n} G_j(x)\right] dx.$$

By recursion, we obtain

$$\int_0^\infty \left[1 - \prod_{j=1}^{n} F_j(x)\right] dx \underset{(\geq)}{\leq} \int_0^\infty \left[1 - \prod_{j=1}^{n} G_j(x)\right] dx. \;\|$$

When $\mu_1 = \mu_i$ ($i = 1, 2, \ldots n$) and F is IFR (DFR), (*b*) becomes

(*b'*) $$\int_0^\infty \left[1 - \prod_{i=1}^{n} F_i(x)\right] dx \underset{(\geq)}{\leq} \mu_1 \sum_{k=1}^{n} \frac{1}{k}.$$

By using Theorem 4.8 and the method of proof in Karlin (1959, Vol. II, pp. 210–214) and Rustagi (1957), we obtain

THEOREM 4.8′. If
 (i) F is IFR, $F(0) = 0$,
 (ii) $\int_0^\infty x\, dF(x) = \mu_1$,
 (iii) $\varphi(x, y)$ is convex (concave) in y and twice differentiable in y,
 (iv) $\partial \varphi(x, y)/\partial y$ is nonincreasing (nondecreasing) in x,
 (v) interchange of integration and partial differentiation with respect to λ is possible in (4.12),

then $$\int_0^\infty \varphi(x, \bar{D}(x))\, dx \underset{(\leq)}{\geq} \int_0^\infty \varphi(x, \bar{F}(x))\, dx \underset{(\leq)}{\geq} \int_0^\infty \varphi(x, e^{-x/\mu_1})\, dx$$

when the indicated integrals exist, where $D(x) = 0$ for x less than μ and 1 for x greater than or equal to μ.

FAILURE DISTRIBUTIONS

Proof. We show the lower bound assuming $\varphi(x, y)$ convex in y and $\frac{\partial}{\partial y} \varphi(x, y)$ nonincreasing in x. Let $\bar{G}_0(x) = e^{-x/\mu_1}$ and define

$$I(\lambda) = \int_0^\infty \varphi(x, \lambda \bar{G}_0(x) + (1 - \lambda) \bar{F}(x))\, dx \qquad (4.12)$$

for $0 \leq \lambda \leq 1$. By using (iii) it is easy to check that $I(\lambda)$ is convex in λ. Hence if $\bar{G}_0(x)$ minimizes (4.12), then $I(\lambda)$ achieves its minimum at $\lambda = 1$. This is possible if and only if

$$I'(\lambda)|_{\lambda=1} \leq 0.$$

Now

$$I'(\lambda)|_{\lambda=1} = \int_0^\infty \frac{\partial}{\partial y} \varphi(x, \bar{G}_0(x))[\bar{G}_0(x) - \bar{F}(x)]\, dx.$$

By (iv) $\partial \varphi(x, y)/\partial y$ is nonincreasing in x. By assumption (iii) $\partial \varphi(x, y)/\partial y$ is nondecreasing in y. But

$$\frac{\partial}{\partial y} \varphi(x, y)\Big|_{y=\bar{G}_0(x)}$$

is nonincreasing in x. Therefore, by Theorem 4.8

$$\int_0^\infty \frac{\partial}{\partial y} \varphi(x, y)\Big|_{y=\bar{G}_0(x)} \bar{F}(x)\, dx \geq \int_0^\infty \frac{\partial}{\partial y} \varphi(x, y)\Big|_{y=\bar{G}_0(x)} \bar{G}_0(x)\, dx,$$

implying $I'(\lambda)|_{\lambda=1} \leq 0$. Hence

$$\int_0^\infty \varphi(x, \bar{F}(x))\, dx \geq \int_0^\infty \varphi(x, \bar{G}_0(x))\, dx.$$

If we assume $\varphi(x, y)$ is concave in y and $\partial \varphi(x, y)/\partial y$ is nondecreasing in x, all inequalities are reversed. The upper bound follows in a similar manner. ∥

Theorem 4.8 is actually a special case in which $\varphi(x, y) = \varphi(x)(1 - y)$ is linear in y.

Theorems 4.8, 4.8′ and Corollaries 4.9, 4.10 may also be obtained as special cases of the corollary on p. 630 of Fan and Lorentz' work (1954).

5. PRESERVATION OF MONOTONE FAILURE RATE

The IFR property has great intuitive appeal for individual units that have been "debugged" and operate in an environment which does not in itself cause units to fail. A natural question is: What structures have the IFR property when their individual components have this property? We shall show that a system consisting of a single IFR unit supported by $n-1$ spares will have this property. In other words, IFR is preserved under convolution. Similarly, the order statistics from an IFR random variable are again IFR. The life of a so-called k-out-of-n structure (see Section 5 of Chapter 7) consisting of n identical components each having an IFR failure distribution corresponds to the kth order statistic. In Chapter 7 the question of preserving increasing failure rate is investigated for more complex structures.

THEOREM 5.1. If F_1 and F_2 are IFR, their convolution H, given by

$$H(t) = \int_{-\infty}^{\infty} F_1(t - x) \, dF_2(x),$$

is also IFR.

Proof. Assume F_1 has density f_1, F_2 has density f_2. For $t_1 < t_2, u_1 < u_2$, form

$$D = |\bar{H}(t_i - u_j)|_{i,j=1,2} = \left| \int F_1(t_i - s) f_2(s - u_j) \, ds \right|$$

$$= \iint_{s_1 < s_2} |F_1(t_i - s_k)| \, |f_2(s_k - u_j)| \, ds_2 \, ds_1$$

by Lemma 1 of Appendix 1. Integrating the inner integral by parts, we obtain

$$D = \iint_{s_1 < s_2} \begin{vmatrix} \bar{F}_1(t_1 - s_1) & f_1(t_1 - s_2) \\ \bar{F}_1(t_2 - s_1) & f_1(t_2 - s_2) \end{vmatrix} \begin{vmatrix} f_2(s_1 - u_1) & f_1(s_1 - u_2) \\ \bar{F}_2(s_2 - u_1) & \bar{F}_2(s_2 - u_2) \end{vmatrix} ds_2 \, ds_1.$$

The sign of the first determinant is the same as that of

$$\frac{f_1(t_2 - s_2)}{\bar{F}_1(t_2 - s_2)} \frac{\bar{F}_1(t_2 - s_2)}{\bar{F}_1(t_2 - s_1)} - \frac{f_1(t_1 - s_2)}{\bar{F}_1(t_1 - s_2)} \frac{\bar{F}_1(t_1 - s_2)}{\bar{F}_1(t_1 - s_1)}$$

assuming nonzero denominators. But

$$\frac{f_1(t_2 - s_2)}{\bar{F}_1(t_2 - s_2)} \geq \frac{f_1(t_1 - s_2)}{\bar{F}_1(t_1 - s_2)}$$

by hypothesis, whereas

$$\frac{\bar{F}_1(t_2 - s_2)}{\bar{F}_1(t_2 - s_1)} \geq \frac{\bar{F}_1(t_1 - s_2)}{\bar{F}_1(t_1 - s_1)}$$

by Theorem 4.1. Thus the first determinant is nonnegative. A similar argument holds for the second determinant, so that $D \geq 0$. But by Theorem 4.1, this implies H is IFR.

FAILURE DISTRIBUTIONS   37

If F and/or G do not have densities, the theorem may be proved in a similar fashion, using limiting arguments. ‖

Theorem 5.1 holds in the discrete case and the proof is similar.

It is of interest to note that the DFR property is *not* preserved under convolution. A counterexample is obtained if F_1 and F_2 have densities f_1 and f_2 such that

$$f_1(x) = f_2(x) = \frac{x^{\alpha-1}e^{-x}}{\Gamma(\alpha)}, \quad x \geq 0$$

with $\frac{1}{2} < \alpha < 1$. It is true, however, that a mixture of DFR distributions is also DFR.

THEOREM 5.2. If $F_i(t)$ is a DFR distribution in t for each $i = 1, 2, \ldots$, $a_i \geq 0$ for each $i = 1, 2, \ldots$, and $\sum_1^\infty a_i = 1$, then

$$G(t) = \sum_{i=1}^\infty a_i F_i(t)$$

is a DFR distribution.

Proof. Suppose each F_i has a differentiable density f_i. Since F_i is DFR, f_i is decreasing. By Schwarz's inequality

$$\sum_{i=1}^\infty a_i \bar{F}_i \sum_{i=1}^\infty a_i(-f_i') \geq \left\{\sum_{i=1}^\infty a_i[\bar{F}_i(-f_i')]^{\frac{1}{2}}\right\}^2.$$

Since $f_i/\bar{F}_i$ is decreasing, $\bar{F}_i(-f_i') \geq f_i^2$. Thus

$$\left\{\sum_{i=1}^\infty a_i[\bar{F}_i(-f_i')]^{\frac{1}{2}}\right\}^2 \geq \left\{\sum_{i=1}^\infty a_i f_i\right\}^2,$$

implying G is DFR.

If any F_i does not have a differentiable density, the result may be obtained by limiting arguments. ‖

Mixtures of IFR distributions are not necessarily IFR. For example, a mixture of two distinct exponentials is not IFR because it is not exponential, and by Theorem 5.2 it is DFR. This theorem together with the test for DFR distributions in Appendix 2 may be used to pick up differences in the parameters of pooled samples each coming from an exponentially distributed population (Proschan, 1963).

Note that in Theorem 5.1 no use was made of the assumption that F_1 and F_2 correspond to positive random variables. A dual to the failure rate is the ratio

$$f(x)/F(x). \tag{5.1}$$

If X is a time variable and time is reversed, then $f(-x)/F(-x)$ becomes the failure rate. Thus a random variable X has increasing failure rate if and only if $-X$ has decreasing (5.1) ratio. Replacing X by $-X$, we obtain from Theorem 4.1(c) that (5.1) is decreasing in x if and only if F is PF_2. The fact that this property is also preserved under convolution will be useful in considering the allocation problem in Chapter 6.

THEOREM 5.3. If F_1 and F_2 are log concave, then

$$H(t) = \int_0^t F_1(t-x)\,dF_2(x)$$

is also log concave.

Proof. Replace X by $-X$ in Theorem 5.1 and use the remark just made concerning the ratio (5.1). ‖

Theorem 5.1 proves that a system consisting of one unit and a spare has increasing failure rate if the components do. As we would expect, the failure rate of the system is everywhere less than the failure rate of either component if both components have IFR failure distributions.

THEOREM 5.4. If F_1 and F_2 are IFR with failure rates $r_1(t)$ and $r_2(t)$ respectively and H denotes their convolution with failure rate $r_h(t)$, then

$$r_h(t) \leq \min\,[r_1(t), r_2(t)].$$

Proof. By definition

$$r_h(t) = \frac{\int_0^t f_1(t-x)f_2(x)\,dx}{\bar{H}(t)}$$

$$\leq r_1(t)\frac{\int_0^t \bar{F}_1(t-x)f_2(x)\,dx}{\bar{H}(t)} = r_1(t),$$

the inequality being clear from

$$\bar{H}(t) = \int_0^\infty \bar{F}_1(t-x)\,dF_2(x).$$

Similarly, $r_h(t) \leq r_2(t)$. ‖

THEOREM 5.5. Assume X is a random variable with distribution F and density f which is IFR. If $X_1, X_2, \ldots, X_n$ are n independent observations on X, the order statistics

$$U_1 < U_2 < \cdots < U_n$$

formed from the X_i's are also IFR.

FAILURE DISTRIBUTIONS   39

Proof. Let H denote the distribution of U_k and $p = \bar{F}(t)$. Then

$$\bar{H}(t) = \sum_{i=k}^{n} \binom{n}{i} [\bar{F}(t)]^i [F(t)]^{n-i}$$

$$= \frac{\Gamma(n+1)}{\Gamma(k)\Gamma(n+1-k)} \int_0^p x^{k-1}(1-x)^{n-k}\, dx,$$

and

$$\frac{\bar{H}(t)}{H'(t)} = \frac{1}{f(t)} \int_0^p \left(\frac{x}{p}\right)^{k-1} \left(\frac{1-x}{1-p}\right)^{n-k} dx.$$

Letting $u = x/p$, we have

$$\frac{\bar{H}(t)}{H'(t)} = \frac{p}{f(t)} \int_0^1 u^{k-1} \left(\frac{1-up}{1-p}\right)^{n-k} du.$$

Because $p/f(t)$ and $(1 - up)/(1 - p)$ are each decreasing in t, so is $\bar{H}(t)/H'(t)$, or $H'(t)/\bar{H}(t)$ is increasing in t. ‖

In particular, Theorem 5.5 tells us that parallel and series structures of identical IFR components are IFR. For series structures, of course, the components do not have to be identical because

$$\log \bar{H}(t) = \log \prod_{i=1}^{n} \bar{F}_i(t) = \sum_{i=1}^{n} \log \bar{F}_i(t)$$

is concave in t if $\log \bar{F}_i(t)$ is concave in t $(i = 1, 2, \ldots, n)$. This implies that H is IFR. Qualitative properties of more complex structures are obtained in Section 6 of Chapter 7.

6. ADDITIONAL INEQUALITIES

The following generalization of Theorem 4.4 is important because it provides additional bounds on the survival probability of an IFR distribution.

THEOREM 6.1. *If F is IFR, $H(x) \geq 0$ is convex increasing, and $H(t) < \int_0^\infty H(x)\, dF(x)$, then*

$$\int_0^\infty H(x)\, dF(x) \leq \int_0^\infty H(x) \frac{L}{t} e^{-Lx/t}\, dx,$$

where $L = -\log \bar{F}(t)$.

Proof. Because $\log \bar{F}(x)$ is concave, there exists z $(0 \leq z < t)$ such that

$$\bar{F}(x) \leq \begin{matrix} 1, & 0 \leq x \leq z \\ \exp\left[\dfrac{x-z}{t-z}(-L)\right], & x \geq z. \end{matrix}$$

H is nonnegative and increasing. Therefore, integrating by parts, we have

$$\int_0^\infty H(x)\,dF(x) \le \frac{L}{1-z}\int_z^\infty H(x)\exp\left[\frac{x-z}{t-z}(-L)\right]dx$$

$$\le L\int_0^\infty H[z(1-y)+ty]e^{-yL}\,dy = \varphi(z).$$

Now φ is convex since H is convex by assumption. Therefore,

$$\int_0^\infty H(x)\,dF(x) \le \varphi(0) \quad\text{or}\quad \varphi(t).$$

But $\varphi(t) = H(t) \ge \int_0^\infty H(x)\,dF(x)$ contradicts the hypothesis of the theorem; therefore,

$$\int_0^\infty H(x)\,dF(x) \le \varphi(0) = \int_0^\infty H(y)\frac{L}{t}e^{-Ly/t}\,dy. \;\|$$

COROLLARY 6.2. If F is IFR, $r \ge 1$, and $\mu_r = \int_0^\infty x^r\,dF(x)$, then

$$\bar F(t) \ge \begin{cases} \exp\left(-\dfrac{t}{\lambda_r^{1/r}}\right), & t < \mu_r^{1/r} \\ 0, & \text{elsewhere} \end{cases}$$

where $\lambda_r = \mu_r/\Gamma(r+1)$.

Proof. Let $H(x) = x^r$ for $r \ge 1$ in Theorem 6.1. $\|$

Analogous bounds can be obtained for $\bar F(t)$ in terms of the moment-generating function or Laplace transform of F at a prescribed point.

Theorem 4.5 can be generalized to

THEOREM 6.3. If F is IFR, $r > 0$, and $\mu_r = \int_0^\infty x^r\,dF(x)$, then

$$\bar F(t) \le \begin{cases} 1, & t < \mu_r^{1/r} \\ e^{-\omega_0 t}, & t \ge \mu_r^{1/r} \end{cases}$$

where ω_0 is the unique solution of $\mu_r = r\int_0^t x^{r-1}e^{-\omega x}\,dx$. The inequality is sharp.

We omit the proof and refer the reader to Barlow and Marshall (1964). A similar generalization of Theorem 4.7 is possible:

THEOREM 6.4. If F is DFR, $r > 0$, and $\mu_r = \int_0^\infty x^r\,dF(x) < \infty$, then

$$\bar F(t) \le \begin{cases} \exp\left(-\dfrac{t}{\lambda_r^{1/r}}\right), & t < r\lambda_r^{1/r} \\ \dfrac{r^r e^{-r}\mu_r}{\Gamma(r+1)t^r}, & t \ge r\lambda_r^{1/r}. \end{cases}$$

FAILURE DISTRIBUTIONS

Upper and lower bounds on $\bar{F}(t)$ when F is IFR and $\mu_1 = 1$, μ_2 are given have been tabulated by Barlow and Marshall (1963). Short tables appear in Appendix 3 as Tables 2 and 3 respectively. The upper bound $U(t)$ given $\mu_1 = 1$ and μ_2 is

$$U(t) = 1, \qquad 0 \leq t \leq 1 - \sqrt{\mu_2 - 1} \qquad (6.1)$$

$$U(t) = e^{-a_1 t}, \qquad 1 - \sqrt{\mu_2 - 1} \leq t \leq -\frac{\log(1-\alpha)}{\alpha} \qquad (6.2)$$

$$U(t) = e^{-b(t-w)}, \qquad t \geq -\frac{\log(1-\alpha)}{\alpha} \qquad (6.3)$$

where α $(0 \leq \alpha \leq 1)$ satisfies

$$\mu_2 = \frac{2}{\alpha}\left[1 + \frac{1-\alpha}{\alpha}\log(1-\alpha)\right]$$

and a_1 is the solution to

$$1 = \frac{1 - e^{-a_1 t}}{a_1} + \frac{e^{-a_1 t}}{a_2}$$

$$\frac{\mu_2}{2} = \frac{1 - (1 + a_1 t)e^{-a_1 t}}{a_1^2} + \frac{(1 + a_2 t)e^{-a_1 t}}{a_2^2}$$

for some $a_2 \geq a_1$ where t satisfies (6.2). Also b and ω satisfy

$$1 = \omega + \frac{1}{b}(1 - e^{-b(t-\omega)})$$

$$\frac{\mu_2}{2} = \frac{\omega^2}{2} + e^{-b\omega}\int_\omega^t x e^{-bx}\,dx$$

where t satisfies (6.3).

The lower bound given two moments is harder to characterize. It is curious, however, that the inequality

$$\bar{F}(\mu_1) \geq e^{-1}$$

cannot be improved even if the first two moments are given and F is IFR.

Table 4 in Appendix 3 tabulates the lower bound on $\bar{F}(t)$ when F is DFR and $\mu_1 = 1$, $\mu_2 = 2(.1)4$.

7. GENERAL FAILURE RATES

The failure rate function provides a natural way of characterizing failure distributions for reliability purposes. Because the failure rate is sometimes not monotone, we shall review some of the general properties of the failure rate function. For example, we can show that $r(t)$ crosses the horizontal line of height $1/\mu_1$ at least once. If the failure rate is increasing, it will cross from below and therefore

$$f(0) = r(0) \leq 1/\mu_1. \tag{7.1}$$

Only in the exponential case is this an equality. However, (7.1) is true under much weaker restrictions. If $r(0) \leq r(t)$ for all t, then $f(0) \leq 1/\mu_1$. To see this explicitly, note that

$$\begin{vmatrix} f(0) & f(t) \\ 1 - F(0) & 1 - F(t) \end{vmatrix} \leq 0$$

by assumption. Integrating on t from 0 to ∞, the assertion becomes clear.

The limiting value of the failure rate plays a significant role in determining the properties of the failure distribution. In order for $F(\infty)$ to equal 1, $\int_0^t r(x)\,dx$ must approach infinity as t approaches infinity. This restricts the rapidity with which $r(x)$ may approach zero. In Theorem 7.1 we prove that all moments of the failure distribution are finite if $\lim_{t \to \infty} r(t) > 0$. It therefore seems reasonable that all failure distributions of interest will have finite moments of all orders. The limit of the failure rate also determines the rate of growth of the distribution.

THEOREM 7.1. If $0 < 1/\alpha \leq r(t) \leq 1/\beta \leq \infty$ for all t, then

$$\mu_s = \int_0^\infty x^s f(x)\,dx < \infty, \quad s > -1 \tag{7.2}$$

$$e^{-t/\beta} \leq \bar{F}(t) \leq e^{-t/\alpha}, \tag{7.3}$$

$$\alpha^{-1} e^{-t/\beta} \leq f(t) \leq \beta^{-1} e^{-t/\alpha}, \tag{7.4}$$

$$\beta^s \leq \lambda_s \leq \alpha^s, \quad s > -1 \tag{7.5}$$

$$\inf_t r(t) \leq 1/\mu_1 \leq \sup_t r(t). \tag{7.6}$$

Proof. Equation (7.3) follows directly from

$$\bar{F}(t) = \exp\left[-\int_0^t r(x)\,dx\right]$$

because by hypothesis

$$-\frac{t}{\beta} \leq -\int_0^t r(x)\,dx \leq -\frac{t}{\alpha};$$

this in turn implies (7.4) and (7.5). Equation (7.2) is a consequence of (7.5). Equation (7.6) is obtained from (7.5) with $s = 1$. ∥

FAILURE DISTRIBUTIONS 43

There is always a corresponding theorem for discrete distributions $\{p_k\}_{k=0}^{\infty}$. In this case, $r(k) = p_k / \sum_{j=k}^{\infty} p_j$.

THEOREM 7.1'. If $0 < (1 + \alpha)^{-1} \leq r(k) \leq (1 + \beta)^{-1} \leq 1, k = 0, 1, \ldots,$ then $\{p_k\}_0^{\infty}$ has finite moments, (7.2)'

$$\left(\frac{\beta}{1+\beta}\right)^k \leq \sum_{j=k}^{\infty} p_j \leq \left(\frac{\alpha}{1+\alpha}\right)^k, \quad k = 0, 1, \ldots \quad (7.3)'$$

$$\frac{1}{1+\alpha}\left(\frac{\beta}{1+\beta}\right)^k \leq p_k \leq \frac{1}{1+\beta}\left(\frac{\alpha}{1+\alpha}\right)^k, \quad (7.4)'$$

$$\beta^m \leq B_m \leq \alpha^m, \quad (7.5)'$$

$$\inf_k r(k) \leq (1 + \mu_1)^{-1} \leq \sup_k r(k), \quad (7.6)'$$

where $B_m = \sum_{j=0}^{\infty} \binom{j}{m} p_j$.

We omit the proof because it is analogous to that of Theorem 7.1. Observe that an equality for $s \neq 0$ in (7.5), say on the left side, implies equality on the left side of (7.3), hence on the left side of (7.5) for all s. Similarly, strict inequality holds in (7.6) except in the exponential case.

The conclusion of (7.2) can be obtained under the considerably weaker condition that $\lim_{t \to \infty} r(t) > 0$ as noted before. For in this case, the truncated density $f_x(t) = f(t + x)/\bar{F}(x)$ $(t > x)$ has failure rate $r_x(t) = r(t + x)$ bounded away from zero for sufficiently large x. Thus f_x (and hence f) has finite moments of all orders.

Many of the moment properties enjoyed by IFR distributions are true under weaker restrictions. For example, suppose F has decreasing mean residual life, that is,

$$\int_t^{\infty} \frac{\bar{F}(x)\, dx}{\bar{F}(t)}$$

is decreasing in t. Let

$$a_{i+j} = \begin{cases} \lambda_{i+j-1}, & i+j > 0 \\ f(0) & i+j = 0 \end{cases}$$

where $\lambda_i = \mu_i/i!$. With these assumptions we can assert that a_{i+k}/a_i is decreasing in i for $i, k = 0, 1, 2, \ldots$ (Barlow, Marshall, Proschan, 1963). In particular, $\{\lambda_i\}_{i=0}^{\infty}$ is a PF_2 sequence, and as a consequence $\lambda_i^{1/i}$ is decreasing in i.

The following theorems provide bounds on the survival probability assuming a variety of conditions on the failure rate function. Proofs of these results can be found in Barlow and Marshall (1964).

THEOREM 7.2. If $r(x) \geq \alpha$ for all $x \geq 0$ and $\int_0^\infty xf(x)\, dx = \mu_1$, then

$$\bar{F}(t) \leq \begin{cases} e^{-\alpha t}, & t \leq -(1/\alpha) \log(1 - \alpha\mu_1) = t_0 \\ \dfrac{\alpha\mu_1 e^{-\alpha t}}{1 - e^{-\alpha t}}, & t \geq t_0 \end{cases}$$

$$\bar{F}(t) \geq \begin{cases} \alpha\mu_1 - 1 + e^{-\alpha t}, & t \leq t_0 \\ 0, & t \geq t_0. \end{cases}$$

These inequalities are sharp.

THEOREM 7.3. If F is IFR, $r(x) \leq \beta$ for all $x \geq 0$, and $\int_0^\infty xf(x)\, dx = \mu_1$, then

$$\bar{F}(t) \leq \begin{cases} 1, & t \leq \mu_1 - 1/\beta \\ \omega_0, & t \geq \mu_1 - 1/\beta \end{cases}$$

where ω_0 is the unique solution of

$$\mu_1 = -\frac{t(1-\omega)}{\log \omega} + \frac{\omega}{\beta},$$

and

$$\bar{F}(t) \geq \begin{cases} e^{-t/\mu_1}, & t \leq \mu_1 \\ e^{-\beta t + \beta \mu_1 - 1}, & t \geq \mu_1. \end{cases}$$

These inequalities are sharp.

THEOREM 7.4. If $r(x) \leq \beta$ for all $x \geq 0$ and $\int_0^\infty xf(x)\, dx = \mu_1$, then

$$\bar{F}(t) \leq \begin{cases} e^{-\beta z_0}, & t > \mu_1 - 1/\beta \\ 1, & t \leq \mu_1 - 1/\beta \end{cases}$$

where z_0 is the unique solution of $(t - z)e^{-\beta z} = \mu_1 - 1/\beta$ satisfying $0 \leq z_0 \leq t$;

$$\bar{F}(t) \geq e^{-\beta t}.$$

These inequalities are sharp.

THEOREM 7.5. If F is IFR, $r(x) \geq \alpha$ for all $x \geq 0$, and $\int_0^\infty xf(x)\, dx = \mu_1$, then

$$\bar{F}(t) \leq \begin{cases} e^{-\alpha t}, & t \leq -(1/\alpha) \log(1 - \alpha\mu_1) = t_0 \\ e^{-\nu t}, & t \geq t_0 \end{cases}$$

FAILURE DISTRIBUTIONS

where y is determined by $(1 - e^{-yt})/y = \mu_1$;

$$\bar{F}(t) \geq \begin{matrix} e^{-t/\mu_1}, & t \leq \mu_1 \\ e^{-(\alpha z+1)}, & \mu_1 < t < t_0 \\ 0, & t \geq t_0 \end{matrix}$$

where z is determined by $1 - \alpha\mu_1 = [1 - \alpha(t - z)]e^{-\alpha z}$. These inequalities are sharp.

Many additional bounds have been obtained on $r(x)$ itself and on the density (Barlow and Marshall, 1964).

CHAPTER 3

Operating Characteristics of Maintenance Policies

1. INTRODUCTION

In many situations, failure of a unit during actual operation is costly or dangerous. If the unit is characterized by a failure rate that increases with age, it may be wise to replace it before it has aged too greatly. In this chapter we shall concentrate on the operating characteristics of some commonly employed replacement policies. By "operating characteristics" we mean the distribution of the number of failures, the distribution of the total number of removals, the expected time to an in-service failure, etc. From a knowledge of the operating characteristics, competing policies may be evaluated with respect to suitably chosen figures of merit, and then compared. The discussion will be almost entirely in terms of replacement of a single unit. Optimal replacement policies for specified cost functions will be considered in the next chapter.

A commonly considered replacement policy is the policy based on age (*age replacement*). Such a policy is in force if a unit is always replaced at the time of failure or T hours after its installation, whichever occurs first; T is a constant unless otherwise specified. If T is a random variable, we shall refer to the policy as a *random age replacement* policy. Under a policy of *block replacement* the unit is replaced at times kT ($k = 1, 2, \ldots$), and at failure. This replacement policy derives its name from the commonly employed practice of replacing a block or group of units in a system at prescribed times kT ($k = 1, 2, \ldots$) independent of the failure history of the system.

Various figures of merit have been proposed for the purpose of evaluating replacement policies. The expected number of failures and the expected number of planned replacements in a specified interval are two quantities which have been studied in detail by B. J. Flehinger (1962a). Another quantity of interest is the probability of no failure in a specified interval. G. Weiss (1956a) was particularly interested in the time to first system failure.

For what component failure distributions would a replacement policy be beneficial? This question has been considered by several authors. G. R. Herd (1955) based his considerations solely on the failure rate. If the failure rate is increasing, replacement should be considered. G. Weiss (1956a), considered an age replacement policy characterized by the time T at which a replacement is to be made if no failure occurs prior to time T. He proposed that an age replacement policy would be beneficial if the expected time to an in-service failure were a decreasing function of T. Another criterion which has been considered is the mean residual life. If the conditional mean life of an item of age t is decreasing in t, an age replacement policy would seem appropriate.

One of the earliest treatments of the replacement problem was by A. J. Lotka (1939). N. R. Campbell (1941), discussed the comparative advantages of replacing a number of street lamps either all at once or as they failed. Clearly the cost per lamp of replacing all lamps at once is less than the cost of replacing each lamp as it fails. The cost of the additional lamps required for preventive maintenance must be balanced against the cost of the additional failures that occur if replacement is postponed. In a series of reports G. Weiss (1956a, 1956b, and 1956c) considered the effects on system reliability and on maintenance costs of both age replacement and random age replacement policies. The operating characteristics of random age replacement policies were also determined by B. J. Flehinger (1962a), and D. M. Brender (1959).

Another variant of the replacement model considers the time for replacement as nonnegligible. This leads to the consideration of a simple two-state stochastic process in which the two states are the "on" state and the "off" state. This model has been studied by L. Takács (1959). An expository paper directed toward reliability applications appeared in *Operations Research* (R. Barlow and L. Hunter, 1961). The following quantities are of particular interest in evaluating competing policies: (*a*) the probability that a unit which is repaired at failure will be on at a specified time; (*b*) the probability that a unit will be operative for t hours or longer during a specified time interval; (*c*) the expected amount of time the unit will be operative during a specified time interval; (*d*) the probability distribution of the number of failures during a specified time interval; (e) the expected number of failures during a specified time interval.

All the replacement problems discussed so far can be treated by the techniques of renewal theory. Before investigating the replacement problem further, it will be necessary to present a summary of the relevant aspects of renewal theory. Since planned replacement is most reasonable for items which wear out with time, we shall emphasize the case in which the underlying failure distribution is IFR.

2. RENEWAL THEORY

Renewal theory has its origins in the study of self-renewing aggregates and especially in actuarial science. We shall now summarize the most important ideas and results of renewal theory for future reference. Most of the general theory may be found in the expository paper by W. L. Smith (1958); the more recent results based on an underlying IFR distribution may be found in R. Barlow, A. Marshall, and F. Proschan (1963) and R. Barlow and F. Proschan (1964).

Definitions

We define a renewal process as a sequence of independent, nonnegative, and identically distributed random variables $X_1, X_2, \ldots$, which are not all zero with probability one. Write F for the distribution of X_1 and $F^{(k)}$ for the distribution of $S_k = X_1 + X_2 + \cdots + X_k$, so that $F^{(k)}$ represents the k-fold convolution of F with itself. We define

$$F^{(0)}(t) = \begin{cases} 1 \text{ for } t \geq 0 \\ 0 \text{ for } t < 0. \end{cases}$$

Renewal theory is primarily concerned with the number of renewals, $N(t)$ in $[0, t]$; more precisely, $N(t)$ is the maximum k for which $S_k \leq t$ subject to the convention $N(t) = 0$ if $X_1 > t$. Clearly,

$$\begin{aligned} P[N(t) = n] &= P[X_1 + X_2 + \cdots + X_n \leq t \\ &\quad \text{and } X_1 + X_2 + \cdots + X_{n+1} > t] \\ &= F^{(n)}(t) - F^{(n+1)}(t). \end{aligned} \quad (2.1)$$

Hence $P[N(t) \geq n] = F^{(n)}(t)$. It is fairly easy to show that $N(t)$ has finite moments of all orders so long as F is not degenerate at the origin. $N(t)$ is called the renewal random variable and the process $\{N(t); t \geq 0\}$ is known as a renewal counting process.

As noted in Theorem 4.4 of Chapter 2, if F is an IFR distribution with mean μ_1, then

$$F^{(n)}(t) \leq 1 - \sum_{j=0}^{n-1} \frac{(t/\mu_1)^j}{j!} e^{-t/\mu_1}$$

OPERATING CHARACTERISTICS OF MAINTENANCE POLICIES 49

for $t < \mu_1$. Therefore

$$P[N(t) \geq n] \leq \sum_{j=n}^{\infty} \frac{(t/\mu_1)^j}{j!} e^{-t/\mu_1}$$

for $t < \mu_1$ when F is IFR with mean μ_1. Thus we have the elementary but important result that, under the IFR assumption, the Poisson distribution provides a conservative estimate of the probability of n or more failures in $[0, t]$ for t less than the mean life of a single component.

Example. The spare parts problem. The following spare parts problem is frequently encountered: How many spare parts should be provided in order to assure with probability α that a system will remain in operation t hours? To make our ideas concrete, suppose that we have one type of tube in n sockets. Suppose, furthermore, that each tube is replaced immediately upon failure. Let $N_j(t_j)$ denote the number of failures (or renewals if the spares are available when needed) occurring in the jth socket before time t_j. Note that the number of spares needed is equal to the number of failures. Here t_j denotes the time that the jth socket is to remain in operation. The t_j may differ because it may not be necessary for all sockets to be in use during the entire operation of the system.

Given α we require the smallest integer N such that

$$P[N_1(t_1) + N_2(t_2) + \cdots + N_n(t_n) \leq N] \geq \alpha.$$

If the sockets are stochastically independent and the life distributions are exponential with parameter λ, that is,

$$F(t) = 1 - e^{-\lambda t},$$

then $N_1(t_1) + N_2(t_2) + \cdots + N_n(t_n)$

is a random variable distributed according to a Poisson law with parameter

$$\theta = \lambda \sum_{j=1}^{n} t_j,$$

and

$$P[N_1(t_1) + \cdots + N_n(t_n) \leq N] = \sum_{j=0}^{N} \frac{\theta^j e^{-\theta}}{j!}$$

(E. Parzen, 1962). If all we can assume is that F is IFR with mean $1/\lambda$, and each $t_i < 1/\lambda$, then

$$P[N_1(t_1) + N_2(t_2) + \cdots + N_n(t_n) \leq N] \geq \sum_{j=0}^{N} \frac{\theta^j}{j!} e^{-\theta}.$$

In this way N, the number of spares to be stocked, can be chosen so that we shall be protected with high probability against a shortage of spares. If different sockets have tubes with different mean lives, we are faced with a multiplicity of ways in which the reliability requirement may be met. If the costs of the spares for the different component types are taken into account, we have the allocation problem solved in Chapter 6.

The renewal function

The renewal function $M(t)$ is defined as the expected number of renewals in $[0, t]$, that is,
$$M(t) = E[N(t)].$$

This function plays a central role in renewal theory and will have many applications in our models. Using (2.1) we have

$$M(t) = E[N(t)] = \sum_{k=1}^{\infty} k P[N(t) = k]$$

or
$$M(t) = \sum_{k=1}^{\infty} F^{(k)}(t). \tag{2.2}$$

Using the fact that
$$F^{(k+1)}(t) = \int_{0-}^{t} F^{(k)}(t - x) \, dF(x)$$

we obtain the *fundamental renewal equation* (2.3):

$$M(t) = F(t) + \sum_{k=1}^{\infty} \int_{0-}^{t} F^{(k)}(t - x) \, dF(x)$$

or
$$M(t) = \int_{0-}^{t} [1 + M(t - x)] \, dF(x). \tag{2.3}$$

If F has a density f, differentiation of (2.3) yields

$$m(t) = \int_{0}^{t} [1 + m(t - x)] f(x) \, dx \tag{2.4}$$

where $m(t) = \dfrac{d}{dt} M(t)$ is known as the *renewal density*. This may also be expressed as

$$m(t) = \sum_{k=1}^{\infty} f^{(k)}(t) \tag{2.5}$$

where $f^{(k)}$ is the k-fold convolution of the density f with itself. From (2.5) it is easy to see that $m(t) \, dt$ is the probability of a renewal occurring in $[t, t + dt]$. This is a very useful probabilistic interpretation.

From (2.3) we obtain the Laplace transform of $M(t)$ as

$$M^*(s) = \int_{0-}^{\infty} e^{-sx} \, dM(x) = \frac{F^*(s)}{1 - F^*(s)} \tag{2.6}$$

where $F^*(s)$ denotes the Laplace transform of F. Since (2.6) implies

OPERATING CHARACTERISTICS OF MAINTENANCE POLICIES   51

$F^*(s) = M^*(s)/[1 + M^*(s)]$ we see that either $M(t)$ or $F(t)$ determines the other.

THEOREM 2.1. If F has mean μ_1, then

$$\frac{N(t)}{t} \to \frac{1}{\mu_1} \quad \text{a.s.†}$$

(We interpret $1/\mu_1 = 0$ whenever $\mu_1 = +\infty$.)

Proof. Suppose $\mu_1 < \infty$. Note that

$$\frac{S_{N(t)}}{N(t)} \leq \frac{t}{N(t)} \leq \frac{S_{N(t)+1}}{[N(t)+1]\frac{N(t)}{N(t)+1}}.$$

Since $N(t) \to \infty$, $S_{N(t)}/N(t) \to \mu_1$ (a.s.) by the strong law of large numbers. Therefore, $t/N(t) \to \mu_1$ (a.s.).

If $\mu_1 = \infty$, we can show that $S_{N(t)}/N(t) \to +\infty$ (a.s.) using a truncation argument. ‖

It is well known that if $F(t) = 1 - e^{-t/\mu_1}$, then

$$M(t) = t/\mu_1$$

for all t and, therefore, $M(t + h) - M(t) = h/\mu_1$. Hence, for the Poisson process, the expected number of renewals in an interval of length h is simply h divided by the mean life. Intuitively, we might expect this to be the case for any renewal process after a long enough time. This is in fact the key theorem of renewal theory for nonlattice random variables. A random variable X is said to be lattice (or periodic) if there exists $h > 0$ such that

$$P[X = nh, n = 0, 1, \text{or} \cdots] = 1.$$

THEOREM 2.2. (Blackwell's theorem). If F is a nonlattice distribution with mean μ_1,

$$\lim_{t \to \infty} [M(t + h) - M(t)] = \frac{h}{\mu_1}.$$

An elementary proof of this fundamental theorem can be found in W. Feller (1961).

It will be convenient to define the *shortage* random variable

$$\delta(t) = t - S_{N(t)} \tag{2.7}$$

and the *excess* random variable

$$\gamma(t) = S_{N(t)+1} - t. \tag{2.8}$$

† The abbreviation "a.s." stands for almost surely; that is, the statement is true with probability one.

Intuitively, $\delta(t)$ is the age of the unit in use at time t and $\gamma(t)$ is the remaining life of the unit in use at time t. Note that

$$P[\delta(t) \leq t] = 1$$

and
$$P[\delta(t) = t] = \bar{F}(t).$$

LEMMA. If $F_1(0) = F_2(0)$, $F_1(x) \geq F_2(x)$ for $0 \leq x \leq t$ and $Q(x) \geq 0$ is nonincreasing on $[0, t]$, then

$$\int_0^t Q(x)\, dF_1(x) \geq \int_0^t Q(x)\, dF_2(x)$$

when the integrals exist.

Proof. Assuming integration by parts is permitted, we obtain

$$\int_0^t Q(x)\, dF_i(x) = Q(x)F_i(x)\Big|_0^t - \int_0^t F_i(x)\, dQ(x)$$

$$= Q(t)F_i(t) - Q(0)F_i(0) + \int_0^t F_i(x)\, d[-Q(x)], \qquad i = 1, 2.$$

Since Q is nonincreasing, $-Q$ is nondecreasing and

$$Q(t)F_1(t) + \int_0^t F_1(x)\, d[-Q(x)] \geq Q(t)F_2(t) + \int_0^t F_2(x)\, d[-Q(x)].$$

If integration by parts is not permissible, we may use limiting arguments to obtain the same result. ‖

The following result is also quite intuitive.

THEOREM 2.3. If F is IFR

$$M(h) \leq M(t + h) - M(t).$$

Proof. Let $F_x(t) = [F(x + t) - F(x)]/\bar{F}(x)$; thus $F_x(t)$ is the failure distribution of an item of age x. Then

$$M(t + h) - M(t) = \int_0^t \int_0^h [1 + M(h - u)]\, dF_x(u)\, d_x P[\delta(t) \leq x]$$

$$\geq \int_0^t \int_0^h [1 + M(h - u)]\, dF(u)\, d_x P[\delta(t) \leq x],$$

since $F_x(t)$ is increasing in x. Therefore

$$M(t + h) - M(t) \geq M(h) \int_0^t d_x P[\delta(t) \leq x] = M(h). \;\|$$

OPERATING CHARACTERISTICS OF MAINTENANCE POLICIES

Thus if F is IFR,

$$M(h) \leq \lim_{t \to \infty} [M(t+h) - M(t)] = \frac{h}{\mu_1} \quad (2.9)$$

for all h by Theorems 2.2 and 2.3. This inequality is in fact true under much weaker assumptions. First, we need a useful result due to Prokhorov and Kolmogorov.

THEOREM 2.4.

$$E[S_{N(t)+1}] = \mu_1[M(t) + 1].$$

An elementary proof can be found in N. L. Johnson (1959).

A stationary renewal process $\{\hat{X}_k\}_{k=1}^{\infty}$ is one for which $\hat{X}_1$ has distribution

$$\hat{F}(t) = \frac{\int_0^t \bar{F}(x)\,dx}{\mu_1}$$

and $\hat{X}_k$ ($k = 2, 3, \ldots$) are independently distributed according to F. Denote a stationary renewal counting process by $\{\hat{N}(t); t \geq 0\}$. It is known (Cox, 1962, p. 46) that $E[\hat{N}(t)] = t/\mu_1$ for this process. By comparing a renewal counting process with its associated stationary process, we can obtain (2.9) assuming only that an aged unit has smaller mean life than a new unit. Recall that the mean life of a unit of age t is

$$\frac{\int_t^{\infty} \bar{F}(x)\,dx}{\bar{F}(t)}.$$

This is also called the mean residual life.

THEOREM 2.5. (a) For any renewal process $M(t) \geq t/\mu_1 - 1$ for all $t \geq 0$.

(b) If $\int_t^{\infty} \frac{\bar{F}(x)\,dx}{\bar{F}(t)} \leq \mu_1$ for all $t \geq 0$, where $\mu_1 = \int_0^{\infty} x\,dF(x)$, then

$$t/\mu_1 - 1 \leq M(t) \leq E[\hat{N}(t)] = t/\mu_1, \quad \text{for } t \geq 0. \quad (2.10)$$

Proof. (a) By Theorem 2.4 we always have

$$E[\gamma(t)] = \mu_1[M(t) + 1] - t \geq 0$$

so that

$$M(t) \geq t/\mu_1 - 1$$

is true for any renewal process.

(b) From the hypothesis it readily follows that

$$\hat{F}(t) = \frac{\int_0^t \bar{F}(x)\,dx}{\mu_1} \geq F(t).$$

Therefore

$$P[\hat{N}(t) \geq n] = \int_0^t F^{(n-1)}(t-x)\,d\hat{F}(x) \geq \int_0^t F^{(n-1)}(t-x)\,dF(x)$$
$$= P[N(t) \geq n].$$

Summing on $n = 1, 2, \ldots$, we obtain (2.10). ∥

Later we shall show that under the IFR assumption we actually have

$$\frac{t}{\mu_1} - 1 \leq \frac{t}{\int_0^t \bar{F}(x)\,dx} - 1 \leq M(t) \leq \frac{t F(t)}{\int_0^t \bar{F}(x)\,dx} \leq \frac{t}{\mu_1}.$$

COROLLARY 2.5. *If F is IFR (DFR) with mean μ_1, then*

(a) $$E[N^k(t)] \underset{(\geq)}{\leq} \sum_{j=0}^{\infty} \frac{j^k (t/\mu_1)^j e^{-t/\mu_1}}{j!}$$

(b) $$\mathrm{Var}\,[N(t)] \underset{(\geq)}{\leq} E[N(t)] \underset{(\geq)}{\leq} t/\mu_1,$$

for $k = 0, 1, \ldots$ and $0 \leq t < \infty$.

Proof. (a) Let $B_m(t) = E\binom{N(t)}{m}$, that is, the mth binomial moment of $N(t)$. It is easily verified, using Laplace transforms, that $B_m(t) = M^{(m)}(t)$ where $M^{(m)}(t)$ denotes the mth convolution of $M(t)$ with itself. Since

$$E[N^k(t)] = \sum_{m=1}^{k+1} S_k^m B_m(t)$$

(Jordan, 1950, p. 168), where S_k^m are the Stirling numbers of the second kind and $S_k^m \geq 0$, (a) follows immediately.

(b) Because $B_2(t) = E\left\{\dfrac{N(t)[N(t)-1]}{2}\right\} = M^{(2)}(t)$, we can compute the variance as

$$\mathrm{Var}\,[N(t)] = 2\int_0^t M(t-x)\,dM(x) + M(t) - [M(t)]^2.$$

To prove (b) we need only show

$$\int_0^t [2M(t-x) - M(t)]\,dM(x) \leq 0.$$

But $M(x) \leq M(t) - M(t-x)$ by Theorem 2.3 implies that we need only show
$$\int_0^t [M(t-x) - M(x)]\, dM(x) \leq 0.$$
Clearly,
$$\int_0^t [M(t-x) - M(x)]\, dM(x)$$
$$= \int_0^{t/2} [M(t-x) - M(x)]\, dM(x) + \int_{t/2}^t [M(t-x) - M(x)]\, dM(x).$$
Let $y = t - x$; then
$$\int_{t/2}^t [M(t-x) - M(x)]\, dM(x) = \int_0^{t/2} [M(t-y) - M(y)]\, dM_y(t-y).$$
Hence we need only show
$$\int_0^{t/2} [M(t-x) - M(x)]\, dM(x)$$
$$\leq \int_0^{t/2} [M(t-x) - M(x)]\, d[M(t) - M(t-x)].$$
This follows immediately, since $M(t-x) - M(x)$ is nonincreasing in x, $[M(t-x) - M(x)] \geq 0$ for $0 \leq x \leq t/2$ and
$$M(x) \leq M(t) - M(t-x)$$
by Theorem 2.3.

All inequalities are reversed if F is DFR. Equality is attained by the Poisson process. ∥

The following so-called *elementary renewal theorem* follows from Blackwell's theorem. Because of its importance, however, we present an elementary proof from Doob.

THEOREM 2.6. (Elementary renewal theorem). If F has mean μ_1, then
$$\lim_{t \to \infty} \frac{M(t)}{t} = \frac{1}{\mu_1}.$$

Proof. By Theorem 2.5, for any renewal process $M(t) \geq t/\mu_1 - 1$ for all t. Therefore
$$\lim_{t \to \infty} \frac{M(t)}{t} \geq \frac{1}{\mu_1}.$$

Define $X_i^{(c)} = \min(X_i, c)$. Then $N^{(c)}(t) \geq N(t)$ and $M^{(c)}(t) \geq M(t)$

where the superscript denotes that the quantity is defined for the renewal process $\{X_i^{(c)}; i \geq 1\}$. By Theorem 2.4

$$\frac{M^{(c)}(t) + 1}{t} = \frac{1}{\mu_1^{(c)}} + \frac{E[\gamma^{(c)}(t)]}{t\mu_1^{(c)}} \leq \frac{1}{\mu_1^{(c)}} + \frac{c}{t\mu_1^{(c)}}$$

since $\gamma(t) \leq c$. Hence

$$\varlimsup_{t\to\infty} \frac{M(t)}{t} \leq \varlimsup_{t\to\infty} \frac{M^{(c)}(t)}{t} = \frac{1}{\mu^{(c)}}, \quad \text{for all } c > 0.$$

As $c \to \infty$, $\mu^{(c)} \to \mu_1$. Therefore, $\varlimsup_{t\to\infty} \frac{M(t)}{t} \leq \frac{1}{\mu_1}$, completing the proof when $\mu_1 < \infty$. If $\mu_1 = +\infty$, then

$$\varlimsup_{t\to\infty} \frac{M(t)}{t} \leq \frac{1}{\mu^{(c)}} \to 0, \quad \text{as } c \to \infty.$$

Therefore $\lim_{t\to\infty} \frac{M(t)}{t} = 0$. ∥

We state the following extension of the elementary renewal theorem without proof.

THEOREM 2.7. If $\mu_2 < \infty$ and F is nonlattice, then

$$M(t) = \frac{t}{\mu_1} + \frac{\mu_2}{2\mu_1^2} - 1 + o(1)$$

as $t \to \infty$.

For a proof of this result, see W. L. Smith (1954).

In general, it is difficult to invert the Laplace transform of $M(t)$ and express it explicitly in terms of the distribution F. $M(t)$ is known for some special but important cases, and we shall list these now for future reference.

Suppose f is the gamma density of order k, that is,

$$f(t) = \frac{\lambda(\lambda t)^{k-1}}{(k-1)!} e^{-\lambda t}.$$

It is easy to verify that f is the convolution of k exponentials with parameter λ. Hence the probability of n renewals in $[0, t]$ for a renewal process defined by f is equal to the probability of either $nk, nk+1, \ldots$ or $nk + k - 1$ events occurring in $[0, t]$ for a Poisson process with parameter λ. Therefore we obtain

$$P[N(t) = n] = \frac{(\lambda t)^{nk}}{(nk)!} e^{-\lambda t} + \frac{(\lambda t)^{nk+1}}{(nk+1)!}$$
$$+ \cdots + \frac{(\lambda t)^{nk+k-1}}{(nk+k-1)!} e^{-\lambda t} \quad (2.11)$$

Let $m(t)$ denote the renewal density for a gamma density of order k. In particular, for $k = 1$ (that is, the exponential density), $m(t) = \lambda$. Since $m(t)\,dt$ is the probability of a renewal in $[t, t + dt]$ we can interpret this probability for the gamma density of order k as

$$m(t)\,dt = \sum_{j=1}^{\infty} \left[\frac{(\lambda t)^{kj-1}}{(kj - 1)!} e^{-\lambda t} \right] \lambda\,dt. \qquad (2.12)$$

The right-hand side is simply the probability of $kj - 1$ events occurring in $[0, t]$ from a Poisson process with parameter λ times the probability of an additional event occurring in $[t, t + dt]$ and summed over all permissible values of j. When $k = 2$

$$m(t) = \lambda/2 - (\lambda/2)e^{-2\lambda t} \qquad (2.13)$$

and
$$M(t) = \lambda t/2 - \tfrac{1}{4} + \tfrac{1}{4}e^{-2\lambda t}. \qquad (2.14)$$

In general, (2.12) can be summed and integrated to obtain

$$M(t) = \frac{\lambda t}{k} + \frac{1}{k} \sum_{j=1}^{k-1} \frac{\theta^j}{1 - \theta^j} [1 - e^{-\lambda t(1 - \theta^j)}]$$

where $\theta = e^{(2\pi i/k)}$. See E. Parzen (1962) for details. $M(t)$ has also been calculated explicitly for a shifted exponential distribution, that is,

$$F(t) = \begin{array}{ll} 0, & t < \beta \\ 1 - e^{-\lambda(t - \beta)}, & t \geq \beta. \end{array}$$

In this case

$$M(t) = \left[\frac{t}{\beta}\right] + \sum_{j=0}^{[t/\beta]} \sum_{k=0}^{j} \lambda^k (t - \beta j)^k \frac{e^{-\lambda(t - \beta j)}}{k!}$$
$$+ \sum_{k=0}^{[t/\beta]} e^{-\lambda(t - \beta k)} \lambda^k \frac{(t - \beta k)^k}{k!} \qquad (2.15)$$

where $[x]$ denotes the greatest integer contained in x (R. Barlow and L. Hunter, 1961, p. 205). We shall also record the value of $M(t)$ for a truncated exponential, that is,

$$F(t) = \begin{array}{ll} 1 - e^{-\lambda t}, & t < T \\ 1, & t \geq T. \end{array}$$

In this case

$$M(t) = \frac{1}{(1 - e^{-T})^2} \{t - (t + T - 1)e^{-T} - e^{-2T}$$
$$- (t + 1 - kT)e^{-kT} + [t + 1 - (k - 1)T]e^{-(k+1)T}\} \qquad (2.16)$$

for $(k - 1)T \leq t < kT$ and $k = 1, 2, \ldots$ (L. Hunter and F. Proschan, 1961, p. 129).

The distribution functions of the shortage random variable $\delta(t)$ (2.7) and the excess random variable $\gamma(t)$ (2.8) can be obtained explicitly in terms of the renewal function.

THEOREM 2.8. For all $t \geq 0$,

(a) $\quad P[\delta(t) \leq x] = \begin{cases} \int_{(t-x)^-}^{t} [\bar{F}(t-u)] \, dM(u), & x < t \\ 1, & x \geq t. \end{cases}$ (2.17)

(b) $\quad P[\gamma(t) \leq x] = \int_{0^-}^{t} [F(t-u+x) - F(t-u)] \, dM(u)$
$\qquad + F(t+x) - F(t).$ (2.18)

Proof.

(a) $\quad P[\delta(t) \leq x] = P[S_{N(t)} \geq t - x]$
$\qquad = \sum_{n=0}^{\infty} P[t - x \leq S_n \leq t, N(t) = n]$
$\qquad = \sum_{n=0}^{\infty} P[t - x \leq S_n \leq t < S_{n+1}]$
$\qquad = \sum_{n=0}^{\infty} \int_{(t-x)^-}^{t} [\bar{F}(t-u)] \, dF^{(n)}(u)$
$\qquad = \int_{(t-x)^-}^{t} [\bar{F}(t-u)] \, dM(u), \quad \text{if } x < t$

and 1 if $x \geq t$.

(b) $\quad P[\gamma(t) \leq x] = \sum_{n=0}^{\infty} P[S_n \leq t < S_{n+1} \leq t - x]$
$\qquad = \sum_{n=0}^{\infty} \int_{0^-}^{t} [F(t+x-u) - F(t-u)] \, dF^{(n)}(u)$
$\qquad = \int_{0^-}^{t} [F(t+x-u) - F(t-u)] \, dM(u)$
$\qquad + F(t+x) - F(t).$ ‖

If F has the density $\lambda e^{-\lambda t}$, it is easy to calculate that

$$P[\gamma(t) > w] = e^{-\lambda w}.$$

This is the only density for which the distribution of $\gamma(t)$ is independent of t. If F has the density $\lambda^2 t e^{-\lambda t}$, then using (2.18) $\gamma(t)$ has the density

$$\lambda^2 x e^{-\lambda x}(\tfrac{1}{2})(1 + e^{-2\lambda t}) + \lambda e^{-\lambda x}(\tfrac{1}{2})(1 - e^{-2\lambda t}).$$

Note that
$$P[\delta(t) \leq w]$$

OPERATING CHARACTERISTICS OF MAINTENANCE POLICIES 59

is the probability that a renewal occurs in $[t - w, t]$ or the probability that the excess at time $t - w$ is less than w, that is, $\gamma(t - w) \leq w$. Hence

$$P[\delta(t) \leq w] = P[\gamma(t - w) \leq w]$$

for $w < t$ and therefore

$$\lim_{t \to \infty} P[\delta(t) \leq w] = \lim_{t \to \infty} P[\gamma(t) \leq w], \qquad (2.19)$$

though the two distributions differ in general for finite t.

By using the distribution of the shortage random variable, bounds can be obtained on the renewal density $m(t)$ in terms of the failure rate $r(t)$. Because $m(t)\, dt$ is the probability of a renewal in $[t, t + dt]$,

$$m(t) = \int_0^t r(x)\, d_x P[\delta(t) \leq x].$$

Therefore

$$\min_{0 \leq x \leq t} r(x) \leq m(t) \leq \max_{0 \leq x \leq t} r(x)$$

and if F is IFR, $m(t) \leq r(t)$. This bound will in general be poor except for relatively small values of t. Of course, in the exponential case we have equality for all $t \geq 0$. To evaluate this bound, note that for the gamma distribution of order 2

$$m(t) = \frac{\lambda}{2} - \frac{\lambda}{2} e^{-2\lambda t} < r(t) = \frac{\lambda^2 t}{1 + \lambda t}, \qquad t > 0.$$

The following fundamental renewal theorem is equivalent to Blackwell's theorem (Theorem 2.2). Because of its useful form we state it separately, however.

THEOREM 2.9. (Fundamental renewal theorem). If g has bounded variation in the interval $[0, \infty]$ and F is nonlattice, we have

$$\lim_{t \to \infty} \int_0^t g(t - u)\, dM(u) = \frac{1}{\mu_1} \int_0^\infty g(u)\, du \qquad (2.20)$$

provided that the integral on the right-hand side exists. An elementary proof of this theorem can be found in L. Takács (1962, p. 227).

COROLLARY 2.10. If F is nonlattice, then

$$\lim_{t \to \infty} P[\delta(t) \geq x] = \lim_{t \to \infty} P[\gamma(t) \geq x] = \frac{\int_x^\infty \bar{F}(u)\, du}{\mu_1} \qquad (2.21)$$

Proof. Using (2.18) we can write

$$P[\gamma(t) \geq x] = \int_0^t \bar{F}(t - u + x)\, dM(u) + \bar{F}(t + x).$$

Letting $g(u) = \bar{F}(u + x)$ in Theorem 2.9, we have

$$\lim_{t \to \infty} \int_0^t \bar{F}(t - u + x)\, dM(u) = \frac{1}{\mu_1} \int_x^\infty \bar{F}(u)\, du.$$

This together with (2.19) completes the proof. ‖

For completeness we shall record the renewal density theorem. A proof can be found in W. L. Smith (1954, p. 42).

THEOREM 2.11. (Renewal density theorem). If f is a density such that $f(x) \to 0$ as $x \to \infty$ and if, for some $p > 1$, $|f(x)|^p$ is integrable, then

$$m(t) \to 1/\mu_1 \quad \text{as } t \to \infty.$$

Certain positivity properties of $N(t)$ for an underlying IFR distribution will be needed to solve the allocation problem in Chapter 6. Hence, we prove that $\log P[N(t) \leq n]$ is a concave sequence in $n = 0, 1, 2, \ldots$ when F is IFR.

THEOREM 2.12. If F is IFR, then $P[N(t) < n]$ is logarithmically concave in $n = 1, 2, \ldots$.

Proof. By Theorems 5.1 and 4.1 of Chapter 2

$$\begin{vmatrix} 1 - F^{(n)}(t) & 1 - F^{(n+1)}(t) \\ 1 - F^{(n)}(t - x) & 1 - F^{(n+1)}(t - x) \end{vmatrix}$$

$$= \int \begin{vmatrix} 1 - F^{(n)}(t) & 1 - F^{(n)}(t - u) \\ 1 - F^{(n)}(t - x) & 1 - F^{(n)}(t - x - u) \end{vmatrix} dF(u) \leq 0.$$

By convolving terms in the last row of the first determinant with F, it will follow that $1 - F^{(n)}(t) = P[N(t) < n]$ is TP_2 in differences of n or equivalently by Theorem 4.1 of Chapter 2 $P[N(t) < n]$ is log concave in $n = 1, 2, \ldots$. ‖

If F has a density f which is PF_2 and $f(x) = 0$ for $x < 0$, we can state a stronger result.

THEOREM 2.13. *If f is a PF_2 density and $f(x) = 0$ for $x < 0$, then $P[N(t) = n]$ is a log concave sequence in $n = 0, 1, \ldots$.*

The proof of this and more general results can be found in S. Karlin and F. Proschan (1960).

3. REPLACEMENT BASED ON AGE

In Section 1 we defined an age replacement policy as one in which a unit is replaced T hours after its installation or at failure, whichever occurs first; T is considered constant. We call this type of replacement policy age replacement. In this situation it is of value to know, for any specified length of time, the distribution and the expected value of the number of planned replacements, the number of failures, and the total number of removals caused by either planned replacement or failure replacement. These quantities will be called the operating characteristics of a specified replacement policy. This kind of information is needed for comparing alternative replacement policies and in determining the number of spares to stock or the budget required to maintain the equipment.

A reasonable assumption in many situations is that the failure mechanism has a wear-out characteristic such that the failure rate is increasing with time. For a specified replacement policy, we are often able to use the operating characteristics which have been calculated for an exponential failure distribution as a *conservative* bound on the true operating characteristics. Some moments of the failure distribution, especially the mean, will be assumed known from prior experience.

Mean time to failure with replacement

Let $\bar{S}_T(t)$ denote the probability that an item does not fail in service before time t; we assume that replacement occurs at failure or T hours after installation, whichever comes first. Then

$$\bar{S}_T(t) = [\bar{F}(T)]^n \bar{F}(t - nT) \tag{3.1}$$

where F is the failure distribution of the unit, $\bar{F} = 1 - F$, and $nT \leq t < (n+1)T$. Assuming that F has a density and differentiating $\bar{S}_T(t)$ with respect to T, we see that

$$\bar{S}_{T_1}(t) \geq \bar{S}_{T_2}(t) \tag{3.2}$$

for all $T_1 \leq T_2$ if and only if F is IFR. In particular,

$$\bar{S}_T(t) \geq \bar{F}(t)$$

for all $T \geq 0$ if F is IFR. This result indicates that age replacement increases survival probability when F is IFR. Of course, this is not true if F has a decreasing failure rate.

The nth moment $E_1^{(n)}(T)$ for the first time to an in-service failure can be calculated from (3.1). In particular,

$$E_1(T) \equiv E_1^{(1)}(T) = \frac{\int_0^T \bar{F}(x)\,dx}{F(T)} \qquad (3.3)$$

and

$$E_1^{(2)}(T) = \frac{2\int_0^T x\bar{F}(x)\,dx}{F(T)} + \frac{2T\bar{F}(T)\int_0^T \bar{F}(x)\,dx}{[F(T)]^2}.$$

The nth moment in terms of Stirling numbers of the second kind has been calculated by G. Weiss (1956a, p. 280).

If F is IFR, it follows from (3.2) that

$$\frac{\int_0^{T_1} \bar{F}(x)\,dx}{F(T_1)} \geq \frac{\int_0^{T_2} \bar{F}(x)\,dx}{F(T_2)}.$$

for $T_1 < T_2$. As we would expect for IFR distributions, the more often we schedule replacement, the longer the mean time to an in-service failure.

Upper bounds on $E_1(T)$ in terms of two moments of F (assumed IFR) can be obtained using the upper bound on $\bar{F}(t)$ tabulated in Table 2 of Appendix 3. Because $\bar{S}_T(t) \geq \bar{F}(t)$ we know, of course, that

$$E_1(T) \geq \mu_1 \qquad (3.4)$$

where $\mu_1 = \int_0^\infty t\,dF(t)$.

In some cases we are more likely to have information concerning a percentile of the failure distribution F rather than a moment. Lower bounds on $E_1(T)$ can be obtained in terms of the pth percentile ξ_p, if $T \leq \xi_p$.

THEOREM 3.1. If F is IFR with density f and mean μ_1, then

(a) $\qquad \mu_1 \leq E_1(T) \leq 1/f(0)$,

(b) $\qquad E_1(T) \geq \xi_p/[-\log(1-p)], \qquad T \leq \xi_p, \qquad (3.5)$

where $\xi_p = \sup\,[t \mid F(t) \leq p] \geq T$.

Proof. (a) The upper bound follows from the fact that $f(0) \leq f(x)/\bar{F}(x)$.
(b) Because $\bar{F}(x) \geq [\bar{F}(\xi_p)]^{x/\xi}$ for $x \leq \xi_p$ by Lemma 4.2 of Chapter 2, we have

$$\int_0^T \bar{F}(x)\,dx \geq \frac{\xi_p\{1 - [\bar{F}(\xi_p)]^{T/\xi_p}\}}{-\log \bar{F}(\xi_p)}$$

and $$F(T) \leq 1 - [\bar{F}(\xi_p)]^{T/\xi_p}$$

implies $$E_1(T) = \frac{\int_0^T \bar{F}(x)\,dx}{F(T)} \geq \frac{\xi_p}{-\log \bar{F}(\xi_p)} \geq \frac{\xi_p}{-\log(1-p)}. \quad \|$$

Note that if ξ_p is the median ($p = 1/2$), then $E_1(T) \geq$ median/log 2 provided $T \leq$ median.

Replacement of redundant structures

Consider a structure consisting of n units, and let $G_n(t)$ denote the failure distribution of the structure with no replacement. Let F denote the common component failure distribution. Suppose that the structure is replaced at age T or at the time of failure, whichever occurs first. Let $E_n(T)$ denote the expected time to the first in-service failure of the structure. Then, since the expected time to first failure is calculated in the same manner as in the case of a single unit, we have

$$E_n(T) = \frac{\int_0^T \bar{G}_n(x)\,dx}{G_n(T)}.$$

[see Equation (3.3)].

If the structure consists of n like IFR units in series, then
$$\bar{G}_n(t) = 1 - [\bar{F}(t)]^n$$
and
$$E_n(\infty) \geq \mu_1/n$$

by Corollary 4.10 of Chapter 2. We know that the failure distribution of a series structure of IFR components is IFR (Section 5 of Chapter 2). Hence $E_n(T)$ is decreasing in T under the IFR assumption and we have

$$E_n(T) \geq \mu_1/n.$$

The bound is sharp. In this case the effect of replacement does not show up in the lower bound. In the parallel case it does.

If a structure consists of n identical units in parallel, each with failure distribution F, then, of course,
$$G_n(t) = [F(t)]^n$$
and
$$\bar{S}_T(t) = \{1 - [F(T)]^n\}^k \{1 - [F(t - kT)]^n\}$$

for $kT \leq t < (k+1)T$. If F is IFR and $T < \mu_1$, then
$$\bar{S}_T(t) \geq [1 - (1 - e^{-T/\mu_1})^n]^k [1 - (1 - e^{-(t-kT)/\mu_1})^n]$$

for $(k-1)T \leq t < kT$. Hence

$$E_n(T) \geq \frac{\int_0^T [1 - (1 - e^{-x/\mu_1})^n]\,dx}{(1 - e^{-T/\mu_1})^n}$$

and computing the right-hand side we have

$$E_n(T) \geq \begin{cases} \mu_1 \sum_1^n \dfrac{(1 - e^{-T/\mu_1})^{n-k}}{k}, & T < \mu_1 \\ \mu_1, & T \geq \mu_1. \end{cases} \quad (3.6)$$

The second lower bound follows from the inequalities

$$\frac{\int_0^T \bar{G}_n(x)\,dx}{G_n(T)} \geq \int_0^\infty \bar{G}_n(x)\,dx \geq \int_0^\infty \bar{F}(x)\,dx = \mu_1.$$

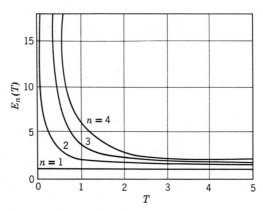

Fig. 3.1. Graph of $E_n(T) = \sum_{k=1}^n \dfrac{(1 - e^{-T})^{n-k}}{k}$.

Because for $T < \mu_1$ equality is attained with the exponential distribution with mean μ_1, and for $T \geq \mu_1$ equality is attained with the degenerate distribution concentrating at μ_1, the bounds in (3.6) are sharp. Figure 3.1 presents comparative graphs of the bound

$$\sum_1^n \frac{(1 - e^{-T})^{n-k}}{k}.$$

This is the first bound of (3.6) when $\mu_1 = 1$.

For $|x| < 1$, by expanding the following right-hand side we see that

$$\sum_1^n \frac{x^k}{k} \geq -\log(1 - x) - \frac{x^n}{1 - x}$$

OPERATING CHARACTERISTICS OF MAINTENANCE POLICIES 65

and hence, letting $x = 1 - e^{-T/\mu_1}$, we obtain

$$\mu_1 \sum_1^n \frac{(1 - e^{-T/\mu_1})^{n-k}}{k} \geq \frac{T}{(1 - e^{-T/\mu_1})^n} - \frac{\mu_1}{e^{-T/\mu_1}} \qquad (3.7)$$

for $T < \mu_1$.

Inequality (3.7) indicates that when a policy of age replacement for a parallel system is adopted, the lower bound on the mean time to system failure goes up approximately exponentially with the number of redundant units.

If there is no replacement, that is, $T = +\infty$, and F is IFR, then for a parallel structure

$$\mu_1 \leq E_n(\infty) \leq \mu_1 \sum_1^n \frac{1}{k}.$$

The upper bound follows from Corollary 4.10 of Chapter 2. Because

$$\mu_1 \sum_1^n \frac{1}{k} \leq \mu_1 (\log n + 1)$$

we see that the improvement in system mean life through redundancy alone can be no more than logarithmic, whereas the improvement through redundancy with replacement is at least exponential.

Clearly we can achieve a specified mean time between failures by either changing T, the time for scheduled replacements, or increasing the number of paralleled units. Using Figure 3.1, we can determine the trade-off between the two methods of increasing mean life.

Exponential bounds on operating characteristics

In this section we assume F is IFR with mean μ_1. In many cases, replacement is scheduled before the estimated unit mean time to failure. In such cases conservative bounds are available on the distribution of the total number of removals (a *removal* refers to either failure or planned replacement) and on the distribution of the number of failures. In particular, let $N_A(t)$ denote the number of removals during $[0, t]$ following a policy of age replacement. If we define

$$G_T(x) = \begin{cases} 1 - e^{-x}, & x < T \\ 1, & x \geq T \end{cases}$$

and

$$F_T(x) = \begin{cases} F(x), & x < T \\ 1, & x \geq T \end{cases}$$

and assume for convenience that $\mu_1 = 1$, then $F_T(x) \leq G_T(x)$ for all x

when $T < 1$ and F is IFR. Under this assumption

$$P[N_A(t) \geq n] = F_T^{(n)}(t) \leq G_T^{(n)}(t)$$

where $F^{(n)}$ denotes the n-fold convolution of F with itself. L. Hunter and F. Proschan (1961) computed $G_T^{(n)}(t)$ to give

$$P[N_A(t) \geq n] \leq \int_0^t \frac{x^{n-1} e^{-x}}{(n-1)!} \, dx$$

$$+ \sum_{j=1}^{\min(k-1,n)} \binom{n}{j} e^{-t} \sum_{m=0}^{j-1} (-1)^m \binom{j-1}{m} \frac{(t-jT)^{n+m-j}}{(n-j+m)!} \quad (3.8)$$

for $(k-1)T \leq t \leq kT$ $(k = 1, 2, \ldots)$.

Hunter and Proschan also computed the expected number of removals in $[0, t]$, $E[N_A(t)]$, for the exponential failure distribution with $\mu_1 = 1$ to give

$$E[N_A(t)] \leq \frac{1}{(1 - e^{-T})^2} \{ t - (t + T - 1)e^{-T} - e^{-2T}$$

$$- (t + 1 - kT)e^{-kT} + [t + 1 - (k-1)T]e^{-(k+1)T} \} \quad (3.9)$$

for $(k-1)T \leq t < kT$ and for $T < 1$. The exponential distribution also provides a bound for the expected time between removals for all T, but in this case a lower bound:

$$\int_0^T \bar{F}(x) \, dx \geq \int_0^T e^{-x/\mu_1} \, dx = (1 - e^{-T/\mu_1})\mu_1 \quad (3.10)$$

assuming F is IFR. Because $\bar{F}(x) \geq e^{-x/\mu_1}$ for $x \leq \mu_1$, the inequality is obvious for $T < \mu_1$. In addition $\bar{F}(x)$ crosses e^{-x/μ_1} exactly once. Hence, if the inequality were ever reversed and strict, we would have $\mu_1 = \int_0^\infty \bar{F}(x) \, dx < \int_0^\infty e^{-x/\mu_1} \, dx = \mu_1$, which is impossible.

Assuming F is IFR and using Theorem 2.5, we find that the expected number of failures in $[0, t]$, $E[N_A^*(t)]$, satisfies

$$\frac{tF(T)}{\int_0^T \bar{F}(x) \, dx} - 1 \leq E[N_A^*(t)] \leq \frac{tF(T)}{\int_0^T \bar{F}(x) \, dx} \leq \frac{t}{\mu_1}$$

for all $T > 0$ and all $t \geq 0$. Similar bounds can be obtained for the expected number of planned replacements.

4. COMPARISON OF AGE AND BLOCK REPLACEMENT POLICIES

Under a policy of block replacement all components of a given type are replaced simultaneously at times kT ($k = 1, 2, \ldots$) independent of the failure history of the system. We also assume that failed components are replaced at failure. These policies have been investigated by R. F. Drenick (1960a), B. J. Flehinger (1962a), and E. L. Welker (1959). They are perhaps more practical than age replacement policies since they do not require the keeping of records on component use. Block replacement policies are commonly used with digital computers and other complex electronic systems.

It will be useful to compare block replacement with age replacement, both using replacement interval T. For example, block replacement is more wasteful since, as we shall show, more unfailed components are removed than under a similar policy based on age. Similarly, the total number of removals for both failed and unfailed components is greater. As might be suspected however, under the IFR assumption the expected number of failures will be less under block replacement. Since exactly $[t/T]$ planned replacements will be made in $[0, t]$ under block replacement, no more than $[t/T]$ can be made under age replacement.

It will be convenient to denote the number of removals under a block policy in $[0, t]$ by $N_B(t)$ and, as in Section 3, the number of removals in $[0, t]$ using age replacement by $N_A(t)$. As we shall show, $N_B(t)$ is stochastically larger than $N_A(t)$.

THEOREM 4.1. $P[N_A(t) \geq n] \leq P[N_B(t) \geq n]$ for $n = 0, 1, 2, \ldots$.

Proof. Let $\{X_k\}_{k=1}^{\infty}$ denote a realization of the lives of successive components. We shall compute what would have occurred under an age and a block replacement policy. Let $T_A^n(T_B^n)$ denote the time of the nth removal under an age (block) replacement policy. Then

$$T_A^n = \min(T_A^{n-1} + T, T_A^{n-1} + X_n)$$
$$T_B^n = \min(T_B^{n-1} + \alpha, T_B^{n-1} + X_n)$$

where α ($0 \leq \alpha \leq T$) is the remaining life to a scheduled replacement. Because initially $T_A^1 = T_B^1$, we have by induction $T_A^n \geq T_B^n$. Thus for any realization $\{X_k\}$ $N_A(t)$ is smaller than $N_B(t)$. ‖

Let $N_A^*(t)[N_B^*(t)]$ denote the number of failures in $[0, t]$ under age (block) replacement at interval T. The next theorem shows that the number of failures per unit time under block replacement at interval T is, in the limit, $M(T)/T$ where $M(T) = \sum_{i=1}^{\infty} F^{(i)}(T)$.

THEOREM 4.2.
$$\lim_{t\to\infty} \frac{N_B^*(t)}{t} = \frac{M(T)}{T} \quad \text{a.s.} \tag{4.1}$$

Proof. Let $N_{Bi}^*(T)$ denote the number of failures in $[(i-1)T, iT]$. Clearly the random variables $N_{Bi}^*(T)$ are independent, identically distributed, and for $kT \leq t < (k+1)T$

$$\sum_{1}^{k} \frac{N_{Bi}^*(T)}{kT[(k+1)/k]} \leq \frac{N_B^*(t)}{t} \leq \sum_{1}^{k+1} \frac{N_{Bi}^*(T)}{(k+1)T[k/(k+1)]}. \tag{4.2}$$

Letting $t \to \infty$, we have $k \to \infty$, and

$$\lim_{t\to\infty} \frac{N_B^*(t)}{t} = \frac{M(T)}{T}$$

by the strong law of large numbers. ∥

From (4.2) we see also that

$$\lim_{t\to\infty} \frac{E[N_B^*(t)]}{t} = \frac{M(t)}{T}. \tag{4.3}$$

Because the number of removals with block replacement is stochastically greater than the number of removals with age replacement, by Theorem 4.1 we see that

$$\lim_{t\to\infty} \frac{E[N_B(t)]}{t} \geq \lim_{t\to\infty} \frac{E[N_A(t)]}{t}. \tag{4.4}$$

But, using (4.3)

$$\lim_{t\to\infty} \frac{E[N_B(t)]}{t} = \frac{M(T)}{T} + \frac{1}{T},$$

and
$$\lim_{t\to\infty} \frac{E[N_A(t)]}{t} = \frac{1}{\int_0^T \bar{F}(x)\,dx}$$

by the elementary renewal theorem, Theorem 2.6, since for age replacement the times between removals constitute a renewal process. Substituting in (4.4) we have

$$\frac{M(T)}{T} + \frac{1}{T} \geq \frac{1}{\int_0^T \bar{F}(x)\,dx}$$

or
$$M(T) \geq \frac{T}{\int_0^T \bar{F}(x)\,dx} - 1 \tag{4.5}$$

OPERATING CHARACTERISTICS OF MAINTENANCE POLICIES 69

which provides a lower bound on the renewal function for all T. This inequality, true for all F, is stronger than the inequality

$$M(T) \geq \frac{T}{\mu_1} - 1$$

of Theorem 2.5 since $\int_0^T \bar{F}(x)\, dx \leq \mu_1$.

To obtain the long-run average of times between failures under a block replacement policy, let $\{Y_i\}$ denote the successive times between failures under a block replacement policy having replacement interval T. Then we have

THEOREM 4.3.

$$\lim_{t \to \infty} \frac{Y_1 + Y_2 + \cdots + Y_{N_B^*(t)}}{N_B^*(t)} = \frac{T}{M(T)} \quad \text{a.s.}$$

Proof. Note that

$$\frac{Y_1 + Y_2 + \cdots + Y_{N_B^*(t)}}{N_B^*(t)} \leq \frac{t}{N_B^*(t)} \leq \frac{Y_1 + Y_2 + \cdots + Y_{N_B^*(t)+1}}{\dfrac{N_B^*(t)}{N_B^*(t) + 1}[N_B^*(t) + 1]}$$

Letting $t \to \infty$ and applying Theorem 4.2, we have the desired result. ∥

The following theorem leads to a useful upper bound on $M(T)$.

THEOREM 4.4. If F is IFR (DFR), then

$$P[N(t) \geq n] \underset{(\leq)}{\geq} P[N_A^*(t) \geq n] \underset{(\leq)}{\geq} P[N_B^*(t) \geq n]$$

for $t \geq 0$, $T > 0$, $n = 0, 1, 2, \ldots$. Equality is attained for the exponential distribution $F(x) = 1 - e^{-x/\mu_1}$ where μ_1 denotes the mean of F.

Proof. Assume F is IFR. First let us suppose $0 \leq t \leq T$, where T is the replacement interval. Let $N(t, x)$ denote the number of renewals in $[0, t]$ in a modified renewal process in which the age of the unit in operation at time 0 is x. Let $N_A^*(t, x)$ denote the number of failures in $[0, t]$ under an age replacement policy in which the age of the unit in operation at time 0 is x. Then we shall show that

$$P[N(t, x) \geq n] \geq P[N_A^*(t, x) \geq n] \geq P[N_B^*(t) \geq n]. \qquad (4.6)$$

For $n = 0$, (4.6) is trivially true. For $n > 0$, we can rewrite (4.6) as

$$\int_0^t F^{(n-1)}(t - u)\, dF_x(u) \geq \int_0^t F^{(n-1)}(t - u)\, dF_x^T(u)$$

$$\geq \int_0^t F^{(n-1)}(t - u)\, dF(u) \qquad (4.6')$$

where $F^{(n)}$ denotes the n-fold convolution of F with itself and

$$F_x(u) = \frac{F(x+u) - F(x)}{\bar{F}(x)}.$$

$F_x^T(u)$ is the distribution of the time to the first failure when the age of the unit in operation at time 0 is x and planned replacement is scheduled for $T - x$, if no failure intervenes. We need specify the distribution $F_x^T(u)$ only on $[0, t]$:

$$F_x^T(u) = \begin{cases} \dfrac{F(x+u) - F(x)}{\bar{F}(x)}, & \text{if } u \leq T - x \\[2mm] \dfrac{F(T) - F(x) + \bar{F}(T)F(u - T + x)}{\bar{F}(x)}, & \text{if } T - x \leq u \leq t. \end{cases}$$

To prove (4.6') we need only show

$$F_x(u) \geq F_x^T(u) \geq F(u), \quad \text{for } 0 \leq u \leq t \tag{4.7}$$

since $F^{(n-1)}(t - u)$ is decreasing in u. For $u \leq T - x$

$$F_x(u) = F_x^T(u) = \frac{F(x+u) - F(x)}{\bar{F}(x)} \geq F(u)$$

because F is IFR. For $T - x \leq u \leq t$

$$\frac{F(x + u - T + T) - F(T)}{\bar{F}(T)} \geq F(x + u - T)$$

implies $\quad F(x + u) \geq F(T) + \bar{F}(T)F(u - T + x)$

and so $\quad F_x(u) = \dfrac{F(x+u) - F(x)}{\bar{F}(x)}$

$$\geq \frac{F(x) - F(T) + \bar{F}(T)F(u - T + x)}{\bar{F}(x)} = F_x^T(u)$$

proves the first inequality in (4.7). Moreover for $T - x < u \leq t$,

$$\frac{\bar{F}(u - T + x)}{\bar{F}(u)}$$

is increasing in u since $\bar{F}$ is PF_2 and since we may assume $x \leq T$. Therefore

$$\frac{\bar{F}(u - T + x)}{\bar{F}(u)} \leq \frac{\bar{F}(x)}{\bar{F}(T)}$$

since $0 \leq u \leq T$. Rearrangement yields

$$\bar{F}(u) \geq \frac{\bar{F}(T)\bar{F}(u - T + x)}{\bar{F}(x)},$$

so that

$$F_x^T(u) = \frac{\bar{F}(x) - \bar{F}(T) + \bar{F}(T)F(u - T + x)}{\bar{F}(x)} \geq F(u)$$

which completes the proof of (4.7). From (4.7) we deduce that for $T \geq x > 0$ and $0 \leq t \leq T$

$$P[N(t, x) \geq n] \geq P[N_A^*(t, x) \geq n] \geq P[N_B^*(t) \geq n],$$

Now suppose $kT < t \leq (k + 1)T$, where $k \geq 1$. The proof proceeds by induction on k. Assume (4.6) is true for $0 \leq t \leq kT$. For $n = 0$, (4.6) is trivially true. For $n > 0$, write

$$P[N(t) \geq n] = \sum_{r=0}^{n} \int_0^T \{P[N(T) = r \mid \delta(T) = x]$$
$$\times P[N(t - T, x) \geq n - r]\} \, d_x P[\delta(T) \leq x]$$

$$P[N_A^*(t) \geq n] = \sum_{r=0}^{n} \int_0^T \{P[N_A^*(T) = r \mid \delta(T) = x]$$
$$\times P[N_A^*(t - T, x) \geq n - r]\} \, d_x P[\delta(T) \leq x]$$

and

$$P[N_B^*(t) \geq n] = \sum_{r=0}^{n} \int_0^T \{P[N_B^*(T) = r \mid \delta(T) = x]$$
$$\times P[N_B^*(t - T) \geq n - r]\} \, d_x P[\delta(T) \leq x]$$

where $\delta(T)$ is a random variable denoting the age of the unit in use at time T. By inductive hypothesis

$$P[N(t - T, x) \geq n - r] \geq P[N_A^*(t - T, x) \geq n - r]$$
$$\geq P[N_B^*(t - T) \geq n - r].$$

In addition,

$$P[N(T) = r \mid \delta(T) = x] = P[N_A^*(T) = r \mid \delta(T) = x]$$
$$= P[N_B^*(T) = r \mid \delta(T) = x]$$

since all three policies coincide on $[0, T]$. Hence (4.6) follows for $kT \leq t \leq (k + 1)T$ for all $k \geq 1$ by the axiom of mathematical induction.

For F (DFR) the proof is similar with the inequalities reversed. ∥

COROLLARY 4.5. If F is IFR,

$$M(t) \leq \frac{t F(t)}{\int_0^t \bar{F}(x) \, dx}, \quad \text{for all } t.$$

Proof. By Theorem 4.4

$$\lim_{t \to \infty} \frac{E[N_B^*(t)]}{t} \leq \lim_{t \to \infty} \frac{E[N_A^*(t)]}{t}. \tag{4.8}$$

By (4.3)

$$\lim_{t \to \infty} \frac{E[N_B^*(t)]}{t} = \frac{M(T)}{T} \tag{4.9}$$

and by (3.3) and the elementary renewal theorem

$$\lim_{t \to \infty} \frac{E[N_A^*(t)]}{t} = \frac{F(T)}{\int_0^T \bar{F}(x)\,dx}. \tag{4.10}$$

Combining (4.8), (4.9), and (4.10) we have the corollary. ‖

From (4.5) and Corollary 4.5 we have the following bounds on the renewal function when F is IFR.

$$\frac{t}{\int_0^t \bar{F}(x)\,dx} - 1 \leq M(t) \leq \frac{tF(t)}{\int_0^t \bar{F}(x)\,dx}. \tag{4.11}$$

These can, of course, be used to bound the cost of block replacement per unit time in the limit.

5. RANDOM REPLACEMENT

It is sometimes impractical to replace a mechanism in a strictly periodic fashion. For example, a given mechanism may have a variable work cycle so that replacement in midcycle is impossible or impractical. In this eventuality the replacement policy would have to be a random one, taking advantage of any free time available for making replacements.

Random age replacement procedures generate several associated renewal processes which can be defined as follows. Suppose first that unit replacements are made only upon failure. Then the times between replacements $\{X_k\}_{k=1}^\infty$ are identically, independently distributed random variables with distribution F and therefore form a renewal process. Next let $\{Y_k\}_{k=1}^\infty$ define a renewal process with corresponding distribution G. This sequence of random variables corresponds to successive planned replacement intervals that do not take into account any actual failures. Define a third associated renewal process $\{Z_k\}$ where $Z_k = \min(X_k, Y_k)$. Then $\{Z_k\}_{k=1}^\infty$ consists of the intervals between successive removals caused by either failures or planned replacements following the replacement procedure defined by G. Let $H(t) = P[Z \leq t]$. Then

$$H(t) = 1 - \bar{G}(t)\bar{F}(t).$$

OPERATING CHARACTERISTICS OF MAINTENANCE POLICIES

The expected time between successive removals is

$$E[Z] = \int_0^\infty \bar{F}(t)\bar{G}(t)\,dt.$$

Another renewal process of interest, $\{V_k\}_{k=1}^\infty$, can be obtained by letting

$$V_k = \begin{cases} 1, & Z_k = X_k \text{ (that is, if the kth removal is due to failure)} \\ 0, & \text{otherwise.} \end{cases}$$

Note that

$$E[V] = P[X \leq Y] = \int_0^\infty F(x)\,dG(x)$$

if F or G is continuous.

Let $N_A(t)$ denote the total number of removals in $[0, t]$. Then

$$\lim_{t \to \infty} \frac{E[N_A(t)]}{t} = \frac{1}{\int_0^\infty \bar{F}(x)\bar{G}(x)\,dx} \quad (5.1)$$

by the elementary renewal theorem.

It is easy to verify that the expected number of actual failures in $[0, t]$ satisfies the following renewal type equation:

$$E[N_A^*(t)] = \int_0^t \{1 + E[N_A^*(t-x)]\}\bar{G}(x)\,dF(x)$$
$$+ \int_0^t E[N_A^*(t-x)]\bar{F}(x)\,dG(x). \quad (5.2)$$

In theory $E[N_A^*(t)]$ can be determined from this equation by Laplace transform methods. The expected number of failures can also be obtained from the distribution of the time between actual failures $\mathscr{F}(t)$. We can in general however, specify this distribution only in terms of the following renewal equation:

$$\mathscr{F}(t) = F(t)\bar{G}(t) + \int_0^t \mathscr{F}(t-x)\bar{F}(x)\,dG(x). \quad (5.3)$$

Useful bounds on $E[N_A^*(t)]$ in terms of $E[N_A(t)]$ can be given:

$$E[V]\{E[N_A(t)] + 1\} - 1 \leq E[N_A^*(t)] \leq E[V]\{E[N_A(t)] + 1\},$$

using the identity

$$E[V_1 + V_2 + \cdots + V_{N_A(t)+1}] = E[V]\{E[N_A(t)] + 1\}.$$

Using (4.11) we have

$$\int_0^\infty F(x)\,dG(x)\,\frac{t}{\int_0^t \bar{H}(x)\,dx} - 1 \leq E[N_A^*(t)]$$

$$\leq \int_0^\infty F(x)\,dG(x)\left[\frac{tH(t)}{\int_0^t \bar{H}(x)\,dx} + 1\right].$$

Similar computations can be made for the number of planned replacements.

As a final result we obtain the limiting *interval reliability* of a unit using a random age replacement policy defined by G. Limiting interval reliability is the limiting probability as $t \to \infty$ that a unit will be operating at time t and will continue to operate for an interval of length x. The unit whose replacement policy we consider may, for example, be a safety device that is used only when an emergency situation arises. It may, however, fail while not being used. These conditions are particularly appropriate to surveillance equipment. We may then ask for the probability that the item will last without replacement for an interval of length x when called into use at time t (called "interval reliability" in Section 1.1). Let $R(x, t)$ denote this probability. Then

$$R(x, t) = \bar{F}(t + x)\bar{G}(t) + \int_0^t \bar{F}(t - u + x)\bar{G}(t - u)\,d_u E[N_A(u)]$$

because $d_u E[N_A(u)]$ is the probability of a removal in $[u, u + du]$. Applying the fundamental renewal theorem (Theorem 2.9), we have

$$\lim_{t \to \infty} R(x, t) = \frac{\int_0^\infty \bar{F}(u + x)\bar{G}(u)\,du}{\int_0^\infty \bar{F}(u)\bar{G}(u)\,du}.$$

B. J. Flehinger (1962a) has generalized all the renewal theory results we have discussed to include more complex replacement policies. All of her results, however, are expressed as solutions of complicated renewal equations and are not available in explicit form.

6. REPAIR OF A SINGLE UNIT

In this section we treat a single unit or component which is repaired upon failure and then returned to operation. Alternatively, we could substitute the time required to make a replacement for repair time. In any event we assume that the unit is as *good as new* after the repair or replacement. Assume that the time to failure is a random variable X with distribution F and that the time to perform a repair is a random variable Y with distribution G. Repair commences immediately upon failure, and, once repaired, the unit is returned to the operating state. The consecutive operating times between failure are assumed to be independently distributed as are the consecutive repair periods. The process of system up and down times can be described by two renewal processes, one superimposed on the other. Perhaps the quantities of greatest interest are the distribution of the number of failures and the distribution of the time that the unit is in a specified state during a given time interval. Also of interest is interval reliability, discussed in the previous section.

G. Weiss (1956d) considered the process with which we are concerned here in connection with the coincidence of periodic pulse trains coming from radar systems. L. Takács (1951) treated the case of exponential failure and general repair distribution as a Type I counter problem. The sojourn time problem has been exhaustively studied by L. Takács (1959). A review article by R. Barlow and L. Hunter (1961) summarizes many of the known results; some of the following results are based on that article.

The distribution of the number of failures

Let H denote the convolution of F and G, that is,

$$H(t) = \int_0^t G(t - x) \, dF(x).$$

It will be convenient to label the operating state by 1 and the failed state by 0. Let $N_{ij}(t)$ denote the number of visits to state j in $(0, t)$, given that the unit enters state i at time $t = 0$. We shall first compute $M_{ij}(t) = E[N_{ij}(t)]$, the expected number of visits to state j in $(0, t)$ if at time 0 the unit enters state i. Note that if the unit begins an "on" period at time 0, the expected number of visits to the "on" state, given that the first failure occurs at time x, is $M_{01}(t - x)$ Therefore

$$M_{11}(t) = \int_0^t M_{01}(t - x) \, dF(x). \tag{6.1}$$

Similarly, if the unit fails at time 0, the expected number of visits to the "on" state, given that the first repair occurs at time x, is $1 + M_{11}(t - x)$. Hence

$$M_{01}(t) = \int_0^t [1 + M_{11}(t - x)] \, dG(x). \tag{6.2}$$

If we know F and G, Equations (6.1) and (6.2) can be used to determine $M_{11}(t)$ and $M_{01}(t)$. One particularly easy way to do this is to take the

Laplace-Stieltjes transform of both sides of these equations. Using an asterisk to denote this transform, that is, $M_{ij}^*(s) = \int_0^\infty e^{-st} dM_{ij}(t)$, we have

$$M_{11}^*(s) = M_{01}^*(s) F^*(s),$$

and $$M_{01}^*(s) = G^*(s) + M_{11}^*(s) G^*(s).$$

These last two equations yield

$$M_{11}^*(s) = \frac{F^*(s) G^*(s)}{1 - F^*(s) G^*(s)},$$

$$M_{01}^*(s) = \frac{G^*(s)}{1 - F^*(s) G^*(s)}. \tag{6.3}$$

In a similar manner we have

$$M_{10}(t) = \int_0^t [1 + M_{00}(t-x)] \, dF(x) \tag{6.4}$$

$$M_{00}(t) = \int_0^t M_{10}(t-x) \, dG(x) \tag{6.5}$$

and again taking Laplace-Stieltjes transforms

$$M_{10}^*(s) = \frac{F^*(s)}{1 - F^*(s) G^*(s)},$$

$$M_{00}^*(s) = \frac{F^*(s) G^*(s)}{1 - F^*(s) G^*(s)}. \tag{6.6}$$

Let $P_{ij}(t)$ denote the probability that the unit is in state j at time t if it starts in state i at $t = 0$. Note that

$$N_{10}(t) - N_{11}(t) = \begin{array}{l} 1, \quad \text{if unit is off at time } t \\ 0, \quad \text{otherwise.} \end{array}$$

Hence $$P_{10}(t) = E[N_{10}(t) - N_{11}(t)] = M_{10}(t) - M_{11}(t), \tag{6.7}$$

and $$P_{11}(t) = 1 - P_{10}(t).$$

Similarly, $$P_{01}(t) = M_{01}(t) - M_{00}(t), \tag{6.8}$$

and $$P_{00}(t) = 1 - P_{01}(t).$$

Letting X with or without subscripts be a random variable denoting the time to failure and similarly letting Y be a random variable denoting the

OPERATING CHARACTERISTICS OF MAINTENANCE POLICIES 77

time for repair, we obtain

$$P[N_{10}(t) = k] = P[X + Y_1 + X_1 + \cdots + Y_{k-1} + X_{k-1} \leq t]$$
$$- P[X + Y_1 + X_1 + \cdots + Y_k + X_k \leq t]$$
$$= F * H^{(k-1)}(t) - F * H^{(k)}(t), \quad k \geq 1$$

and
$$P[N_{10}(t) = 0] = 1 - F(t),$$

where the asterisk denotes pairwise convolution and $H^{(k)}(t)$ means that H is convoluted with itself k times.

The Laplace transform of the distribution of the number of failures can be obtained in the following way. Let $W(t, n)$ be the probability of n or less failures in time t starting with the unit on. Then

$$W(t, n) = \sum_{k=0}^{n} P[N_{10}(t) = k]$$
$$= 1 - F(t) + \sum_{k=1}^{n} [F * H^{(k-1)}(t) - F * H^{(k)}(t)] = 1 - F * H^{(n)}(t) \quad (6.9)$$

where
$$H^{(0)}(t) = \begin{array}{l} 0, \quad \text{if } t < 0 \\ 1, \quad \text{if } t \geq 0. \end{array}$$

Then
$$W^*(s, n) = \int_0^\infty e^{-st} d_t W(t, n) = -F^*(s)[H^*(s)]^n. \quad (6.10)$$

In general, these Laplace-Stieltjes transforms are difficult to invert. By using Theorem 2.7, however, we can state the following asymptotic formulas describing behavior for large t, assuming F or G nonlattice.

$$M_{10}(t) = \frac{t}{\mu_1 + \nu_1} - \frac{\mu_1}{\mu_1 + \nu_1} + \frac{l^{(2)}}{2(\mu_1 + \nu_1)^2} + o(1) \quad (6.11)$$

$$M_{11}(t) = \frac{t}{\mu_1 + \nu_1} - 1 + \frac{l^{(2)}}{2(\mu_1 + \nu_1)^2} + o(1) \quad (6.12)$$

and
$$P_{10}(t) = M_{10}(t) - M_{11}(t) = \frac{\nu_1}{\mu_1 + \nu_1} + o(1) \quad (6.13)$$

where μ_1 and ν_1 are the means of F and G, respectively, and $l^{(2)}$ is the second moment of H.

Example 1. Exponential failure and exponential repair time. Suppose $F(t) = 1 - e^{-at}$, and $G(t) = 1 - e^{-bt}$. Then

$$F^*(s) = \frac{a}{s + a}, \quad G^*(s) = \frac{b}{s + b}.$$

The Laplace-Stieltjes transform of the expected number of failures is by (6.6)

$$M_{10}^*(s) = \frac{a(s+b)}{s^2 + (a+b)s}.$$

Inverting, we have

$$M_{10}(t) = \frac{a^2}{(a+b)^2} + \frac{abt}{a+b} - \frac{a^2 e^{-(a+b)t}}{(a+b)^2}.$$

In a similar manner, we compute

$$M_{11}(t) = -\frac{ab}{(a+b)^2} + \frac{abt}{a+b} + \frac{abe^{-(a+b)t}}{(a+b)^2}.$$

Thus

$$P_{10}(t) = M_{10}(t) - M_{11}(t) = \frac{a}{a+b} - \frac{ae^{-(a+b)t}}{a+b},$$

and

$$P_{11}(t) = \frac{b}{a+b} + \frac{ae^{-(a+b)t}}{a+b}.$$

Upon inversion of (6.10), we obtain

$$W(t, n) = \sum_{j=1}^{n+1} \frac{A_j t^{j-1}}{(j-1)!} e^{-at} + \sum_{j=1}^{n} \frac{B_j t^{j-1}}{(j-1)!} e^{-bt},$$

where

$$A_j = a^{j-1} + \sum_{k=1}^{n} (-1)^{n+1-j} \binom{n+k-j}{k-1} \frac{b^{k-1} a^{n+1}}{(b-a)^{n+k-j+1}},$$

and

$$B_j = b^{j-1} + \sum_{k=1}^{n+1} (-1)^k \binom{n+k-j+1}{k-1} \frac{a^{k-1} b^{n+2k-2j}}{(b-a)^{n+k-j}}.$$

Example 2. Exponential failure and constant repair time. Suppose again that the failure distribution is exponential, but that the time for repair is constant, say β. Then

$$F^*(s) = \frac{a}{a+s}, \quad G^*(s) = e^{-\beta s}.$$

The mean number of failures in $(0, t)$ becomes

$$M_{10}(t) = \left[\frac{t}{\beta}\right] + 1 + \sum_{j=0}^{[t/\beta]} \sum_{k=0}^{j} \frac{a^k (t-\beta j)^k e^{-a(t-\beta j)}}{k!},$$

and

$$M_{11}(t) = \left[\frac{t}{\beta}\right] + \sum_{j=0}^{[t/\beta]} \sum_{k=0}^{j} \frac{a^k (t-\beta j)^k e^{-a(t-\beta j)}}{k!}$$

$$+ \sum_{k=0}^{[t/\beta]} \frac{e^{-a(t-\beta k)} a^k (t-\beta k)^k}{k!},$$

OPERATING CHARACTERISTICS OF MAINTENANCE POLICIES 79

where $[t/\beta]$ denotes the greatest integer contained in t/β. Hence

$$P_{10}(t) = 1 - \sum_{j=0}^{[t/\beta]} \frac{a^j(t - \beta j)^j e^{-a(t-\beta j)}}{j!},$$

and
$$P_{11}(t) = 1 - P_{10}(t).$$

The distribution of the number of failures can be obtained as

$$W(t, n) = \begin{cases} \sum_{j=1}^{n+1} \frac{a^{j-1}(t - n\beta)^{j-1} e^{-a(t-n\beta)}}{(j-1)!}, & n\beta \leq t \\ 1, & n\beta > t. \end{cases}$$

The distribution of unit down time

Let $\gamma(t)$ denote the amount of time the unit is down during $(0, t)$. That is, $\gamma(t)$ is the amount of time spent in repairing the unit in time t. Put

$$\Omega(t, x) = P[\gamma(t) \leq x].$$

L. Takács (1957a) showed that the distribution function for $\gamma(t)$ is

$$\Omega(t, x) = \sum_{n=0}^{\infty} G^{(n)}(x)[F^{(n)}(t - x) - F^{(n+1)}(t - x)]. \tag{6.14}$$

Takács also proved the following important theorem.

THEOREM 6.1. Let $\mu_1 = EX$, $\nu_1 = EY$, $\sigma_1^2 =$ variance of X, $\sigma_2^2 =$ variance of Y. If $\sigma_1^2 < \infty$, $\sigma_2^2 < \infty$, then

$$\lim_{t \to \infty} P\left[\frac{\gamma(t) - \nu_1 t/(\mu_1 + \nu_1)}{\sqrt{(\nu_1^2 \sigma_1^2 + \mu_1^2 \sigma_2^2)t/(\mu_1 + \nu_1)^3}} \leq x\right] = \frac{1}{\sqrt{2\pi}} \int_{-\infty}^{x} e^{-u^2/2} \, du.$$

Example 3. Asymptotic distribution of down time. Suppose we wish to compute the probability that the unit is down more than 24 hours in 10,000 hours of operation. The failure and repair distributions are unknown, but from sample data the following estimates of means and variances are made:

$$\mu_1 = 1000 \text{ hours}, \quad \sigma_1^2 = 100{,}000$$

$$\nu_1 = 2 \text{ hours} \quad \sigma_2^2 = 4.$$

Appealing to Theorem 6.1, we know that, for $t = 10{,}000$,

$$\frac{\gamma(t) - \nu_1 t/(\mu_1 + \nu_1)}{(\nu_1^2 \sigma_1^2 + \mu_1^2 \sigma_2^2)t/(\mu_1 + \nu_1)^3} = \frac{\gamma(10{,}000) - 19.96}{6.66}$$

is approximately normally distributed with mean 0 and variance 1. Hence

$$P[\gamma(10,000) \geq 24] = P\left[\frac{\gamma - 19.96}{6.66} \geq \frac{24 - 19.96}{6.66}\right]$$

$$\approx \frac{1}{\sqrt{2\pi}} \int_{.6}^{\infty} e^{-u^2/2}\, du \approx .28.$$

Thus the probability that the unit is down more than 24 hours in 10,000 hours is about .28.

Example 4. Exponential failure and exponential repair time. It is easy to work out the exact distribution for the down time, given by Equation (6.14), when $F(t)$ and $G(t)$ are both exponentials:

$$F(t) = 1 - e^{-at}, \quad G(t) = 1 - e^{-bt}.$$

Then
$$F^{(n)}(t-x) - F^{(n+1)}(t-x) = e^{-a(t-x)}\frac{[a(t-x)]^n}{n!},$$

and
$$\Omega(u+x, x) = \sum_{n=0}^{n} e^{-au}\frac{(au)^n}{n!} G^{(n)}(x).$$

The Laplace-Stieltjes transform is

$$\int_0^{\infty} e^{-sx}\, d_x\Omega(u+x, x) = e^{-au[1-R(s)]}$$

where
$$R(s) = \int_0^{\infty} e^{-sx}\, dG(x) = \frac{b}{b+s}.$$

So by inversion

$$\Omega(u+x, x) = e^{-au}\left[1 + \sqrt{abu}\int_0^x e^{-by}y^{-1/2}I_1(2\sqrt{abuy})\, dy\right],$$

where $I_1(x)$ is the Bessel function of order 1 for the imaginary argument defined by

$$I_1(x) = \sum_{j=0}^{\infty}\frac{(\tfrac{1}{2}x)^{2j+1}}{j!(j+1)!}.$$

By substituting $u = t - x$ we obtain

$$\Omega(t, x) = e^{-a(t-x)}\left[1 + \sqrt{ab(t-x)}\int_0^x e^{-by}y^{-1/2}I_1(2\sqrt{ab(t-x)y})\, dy\right].$$

The expected unit "on" time

The expected time that the unit is on in $(0, T)$ is

$$\int_0^T P_{11}(x)\, dx.$$

OPERATING CHARACTERISTICS OF MAINTENANCE POLICIES

To see this let
$$Z(t) = \begin{cases} 1, & \text{if the unit is on at time } t \\ 0, & \text{otherwise.} \end{cases}$$

Let L equal the length of time the unit is in the on state in $(0, T)$. Then because the unit starts operating at $t = 0$,

$$E[Z(t)] = P_{11}(t)$$

and using the Fubini theorem,

$$\int_0^T P_{11}(t)\,dt = \int_0^T E[Z(t)]\,dt = E\left[\int_0^T Z(t)\,dt\right] = E[L].$$

From Example 1, we see that when $F(t) = 1 - e^{-at}$ and $G(t) = 1 - e^{-bt}$

$$\int_0^T P_{11}(x)\,dx = \frac{bT}{a+b} + \frac{a}{(a+b)^2}(1 - e^{-(a+b)T})$$

and in this case, $bT/(a+b)$ is a conservative estimate of average unit on time. We shall show that this is in fact the case whenever F is IFR.

In Barlow (1962b), it is shown that

$$\lim_{T \to \infty} \int_0^T \left[P_{11}(x) - \frac{\mu_1}{\mu_1 + \nu_1}\right] dx = \frac{\mu_1 l^{(2)} - \mu_2(\mu_1 + \nu_1)}{2(\mu_1 + \nu_1)^2} \quad (6.15)$$

where μ_1, μ_2 are the mean and second moment of F, ν_1 is the mean of G, and $l^{(2)}$ is the second moment of the convolution of F and G.

From (6.15) we see that $T\mu_1/(\mu_1 + \nu_1)$ is a good estimate for the expected on time in $[0, T]$.

THEOREM 6.2. If F is IFR, then

$$\lim_{T \to \infty} \int_0^T \left[P_{11}(x) - \frac{\mu_1}{\mu_1 + \nu_1}\right] dx \geq 0.$$

Proof. Assume F is IFR. By (6.15) we need only show

$$\begin{vmatrix} \mu_1 & \mu_1 + \nu_1 \\ \mu_2 & l^{(2)} \end{vmatrix} \geq 0$$

or

$$\Delta = \begin{vmatrix} \int_0^\infty \bar{F}(t)\,dt & \int_0^\infty \int_0^\infty \bar{F}(t-\theta)\,dG(\theta)\,dt \\ \int_0^\infty t\bar{F}(t)\,dt & \int_0^\infty t\int_0^\infty \bar{F}(t-\theta)\,dG(\theta)\,dt \end{vmatrix} \geq 0.$$

Using Lemma 1 of Appendix 1 we have

$$\Delta = \iint_{t_1 < t_2} \begin{vmatrix} 1 & 1 \\ t_1 & t_2 \end{vmatrix} \int_0^\infty \begin{vmatrix} \bar{F}(t_1 - 0) & \bar{F}(t_1 - \theta) \\ \bar{F}(t_2 - 0) & \bar{F}(t_2 - \theta) \end{vmatrix} dG(\theta) \, dt_1 \, dt_2$$

where, of course, $\bar{F}(x) = 1$ when $x < 0$. Using part (c) of Theorem 4.1 of Chapter 2, we see that the determinants are clearly nonnegative and, therefore, $\Delta \geq 0$. ∥

Therefore $T\mu_1/(\mu_1 + \nu_1)$ underestimates the average unit on time for large T when F is IFR.

Interval reliability

We may ask for the probability that a unit will survive for an interval of length x without repair when called into use at time t. This quantity is known as the interval reliability of a unit and was defined in Section 5. Let $R(x, t)$ denote this probability. Then

$$R(x, t) = \bar{F}(t + x) + \int_0^t \bar{F}(t - y + x) \, dM_{11}(y),$$

since $dM_{11}(y)$ is the probability of a completed repair occurring at time y. The limit of this quantity as t approaches infinity follows from Theorem 2.9, that is,

$$\lim_{t \to \infty} R(x, t) = \frac{\int_x^\infty \bar{F}(y) \, dy}{\mu_1 + \nu_1}$$

assuming F or G is nonlattice. It is always true that

$$\frac{\int_x^\infty \bar{F}(u) \, du}{\mu_1 + \nu_1} \geq \frac{\mu_1 - x}{\mu_1 + \nu_1}.$$

If F is IFR, then we have

$$\frac{\mu_1 - x}{\mu_1 + \nu_1} \leq \frac{\int_x^\infty \bar{F}(u) \, du}{\mu_1 + \nu_1} \leq \frac{\mu_1 e^{-x/\mu_1}}{\mu_1 + \nu_1}$$

because by the remark preceding Lemma 4.2 of Chapter 2

$$\int_x^\infty \bar{F}(u) \, du \leq \int_x^\infty e^{-u/\mu_1} \, du$$

for all x.

We may also ask for the probability that the unit will be repaired within a sufficiently short interval if it is not operating at a specified time. We

OPERATING CHARACTERISTICS OF MAINTENANCE POLICIES   83

call this quantity the servicing reliability and use the notation $S(x, t)$. Thus $S(x, t)$ is the probability either that the unit is operating at time t or, if it is not operating, that it will be repaired within an interval of length x. Reasoning as before we obtain

$$S(x, t) = 1 - \int_0^t \bar{G}(t - y + x) \, dM_{10}(y).$$

Again, applying the fundamental renewal theorem, Theorem 2.9, we obtain

$$\lim_{t \to \infty} S(x, t) = 1 - \frac{\int_x^\infty \bar{G}(u) \, du}{\mu_1 + \nu_1},$$

assuming F or G is nonlattice.

CHAPTER 4

Optimum Maintenance Policies

1. INTRODUCTION

In Chapter 3 various maintenance policies were studied with a view toward obtaining information concerning their basic stochastic characteristics, such as the distribution of the number of failures, the distribution of the total number of removals, the expected time to an in-service failure, etc. In the present chapter we attempt to find or characterize *optimum* maintenance policies; that is, we seek the member of a specific class of maintenance policies that minimizes total cost, maximizes availability, or in general attains the best value of the prescribed objective function.

Several broad classes of maintenance policies are considered. In Section 2 policies are treated governing the scheduling of replacement of equipment so as to forestall failure during operation. In Section 3 inspection policies typically aimed at achieving maximum operational readiness are treated; in these so-called "preparedness" models, failure is known only through checking. In the following chapter more complicated formulations are considered in which decisions concerning replacement, repair, and inspection are made at each successive step; the model is Markovian in that the decision depends only on information concerning the present state of the system and not on its past history.

The replacement policies of Section 2 are further classified as policies of age replacement and of block replacement, and as policies of periodic replacement (with minimal repair at time of failure) and of sequential replacement. Age and block replacement were defined, studied, and

OPTIMUM MAINTENANCE POLICIES 85

compared in Chapter 3, and periodic and sequential replacement are introduced in the present chapter. In the finite time span replacement models we try to minimize expected cost $C(t)$ experienced during $[0, t]$, where cost may be computed in dollars, time, or some appropriate combination.

For an infinite time span, an appropriate objective function is expected cost per unit of time, expressed as

$$\lim_{t \to \infty} \frac{C(t)}{t}. \qquad (1.1)$$

As is generally true, the optimum replacement policy is more readily obtained for an infinite time span than for a finite time span.

2. REPLACEMENT POLICIES

In this section we consider the problem of specifying a replacement policy which balances the cost of failures of a unit during operation against the cost of planned replacements; some of the material is based on Barlow and Proschan (1962). In all the replacement models to be considered, a cost c_1 is suffered for each failed item which is replaced; this includes all costs resulting from the failure and its replacement. We assume that failures are instantly detected and replaced. A cost $c_2 < c_1$ is suffered for each nonfailed item which is exchanged. Letting $N_1(t)$ denote the number of failures during $[0, t]$ and $N_2(t)$ denote the number of exchanges of nonfailed items during $[0, t]$, we may express the expected cost during $[0, t]$ as

$$C(t) = c_1 E N_1(t) + c_2 E N_2(t) \qquad (2.1)$$

We shall seek the policy minimizing $C(t)$ for a finite time span or minimizing $\lim_{t \to \infty} C(t)/t$ for an infinite time span within each of the following classes: (1) age replacement (Section 2.1), (2) block replacement (Section 2.2), (3) periodic replacement with minimal repair at failure (Section 2.3), and finally, (4) sequential replacement (Section 2.4). We can also interpret c_1 as the mean time to replace a failed component and c_2 as the mean time to replace a nonfailed component. Then $C(t)$ becomes the expected down time in $[0, t]$. The replacement policy minimizing $C(t)$ will then maximize limiting availability.

It is apparent that none of the replacement policies mentioned above is appropriate if the underlying failure distribution is DFR, for in this case used items tend to have a longer remaining life than their replacements.

2.1 Age replacement

Recall that an age replacement policy is in effect if replacement occurs at failure or at age T, whichever occurs first. Unless otherwise specified,

T is taken to be a constant. If T is a random variable, independently chosen from a fixed distribution for each scheduled replacement, we call the policy "random age replacement" (see Section 5 of Chapter 3).

Age replacement for an infinite time span seems to have received the most attention in the literature. For such policies Morse (1958) shows how to determine the replacement interval minimizing expected cost per unit of time, Equation (1.1), when such an optimum interval exists. The derivation of an optimum age replacement interval corresponding to a given finite time span is basically a much more difficult problem. Barlow and Proschan (1962) prove the existence of such an optimal policy; they illustrate its calculation for a specific case. Derman and Sacks (1960) obtain the optimal replacement policy for a piece of equipment in which the decision to replace depends on the observed state of equipment deterioration at specified points in time.

Infinite Time Span. We need only consider nonrandom age replacement in seeking the optimum policy for an infinite time span, as shown in

THEOREM 2.1. Assume the underlying failure distribution F is continuous. Then the optimum age replacement policy is nonrandom for an infinite time span.

Proof. Let $A(G)$ denote the average cost per unit of time over an indefinitely long period if we follow a random age replacement policy with replacement interval T governed by distribution G. Note that $T = \infty$ corresponds to replacement at failure only. By definition

$$A(G) = \lim_{t \to \infty} \left[c_1 \frac{EN_1(t)}{t} + c_2 \frac{EN_2(t)}{t} \right].$$

Using Theorem 2.6 of Chapter 3, we may show

$$A(G) = \frac{c_1 \int_0^\infty F(x)\, dG(x) + c_2 \int_0^\infty G(x)\, dF(x)}{\int_0^\infty \bar{G}(x)\bar{F}(x)\, dx}. \tag{2.2}$$

(More explicitly (2.2) follows from results 7 and 8, p. 75, of Barlow and Proschan, 1962.) We can write (2.2) as

$$A(G) = \frac{\int_0^\infty Q(x)\, dG(x)}{\int_0^\infty S(x)\, dG(x)},$$

where
$$Q(x) = c_1 F(x) + c_2 \bar{F}(x),$$

and
$$S(x) = \int_0^x y\, dF(y) + x \int_x^\infty dF(y).$$

Suppose now we find a minimum value of $Q(x)/S(x)$ as x ranges over $[0, \infty]$; such a minimum exists since $Q(x)/S(x)$ is continuous for $x > 0$. Let x_0 (possibly infinite) be a minimizing x. Then since

$$\frac{Q(x)}{S(x)} \geq \frac{Q(x_0)}{S(x_0)}, \quad \left(\frac{Q(x)}{S(x)} \geq \lim_{x \to \infty} \frac{Q(x)}{S(x)} \text{ if } x_0 = \infty\right),$$

it follows that
$$\int_0^\infty Q(x)\, dG(x) \geq \frac{Q(x_0)}{S(x_0)} \int_0^\infty S(x)\, dG(x),$$

so that
$$A(G) \geq \frac{Q(x_0)}{S(x_0)} = A(G_0),$$

where G_0 is the degenerate distribution placing unit mass at x_0; if x_0 is infinite, G_0 corresponds to replacement only at failure. Thus the optimum age replacement policy is nonrandom. ‖

This argument is due to Prof. S. Karlin.

To find the optimum replacement interval if one exists, we seek the value x_0 of Theorem 2.1 which minimizes $Q(x)/S(x)$ for x in $[0, \infty]$. If the failure distribution F has a density f, a necessary condition that $x_0 < \infty$ minimize

$$\frac{Q(x)}{S(x)} = \frac{c_1 F(x) + c_2 \bar{F}(x)}{\int_0^x \bar{F}(u)\, du}$$

is obtained by setting the derivative of $Q(x)/S(x)$ equal to 0:

$$r(x) \int_0^x \bar{F}(t)\, dt - F(x) = \frac{c_2}{c_1 - c_2}, \qquad (2.3)$$

where $r(x)$ represents the failure rate $f(x)/\bar{F}(x)$. If we assume further that $r(x)$ is continuous and increasing, the left side of (2.3) is continuous and increasing, and an optimum policy x_0 exists, perhaps infinite. If F has a continuous density and no solution to (2.3) exists, then since $Q(x)/S(x) \to \infty$ as $x \to 0$ it follows that $Q(x)/S(x)$ is decreasing and so $x_0 = \infty$, that is, a policy of replacement only at failure is optimal. Finally, note that a solution to (2.3) is unique and finite if $r(x)$ is continuous and strictly increasing to infinity. Recall that for the truncated normal distribution and the Weibull with parameter $\alpha > 1$ (see Chapter 2, Section 2), $r(x)$ is continuous and strictly increasing to infinity.

Let

$$L(x) = \frac{c_1 F(x) + c_2 \bar{F}(x)}{\int_0^x \bar{F}(u)\,du}.$$

Then $L(0) = \infty$ and $L(\infty) = c_1/\mu_1$. $dL/dx = 0$ implies

$$r(x) \int_0^x \bar{F}(t)\,dt - F(x) = \frac{c_2}{c_1 - c_2}.$$

When F is IFR the left-hand side is increasing. Hence $L(x)$ has at most one minimum. Thus $L(x)$ looks like the curve in Figure 2.1a or b.

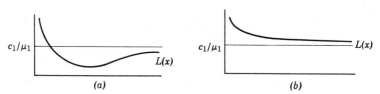

Fig. 2.1

Now

$$L(x) = \frac{c_2 + (c_1 - c_2)F(x)}{\int_0^x \bar{F}(u)\,du} \geq \frac{c_2}{x} \quad \text{for } c_1 \geq c_2.$$

Thus $L(x)$ must cross c_1/μ to the right of where c_2/x crosses c_1/μ_1, if it crosses at all. But the latter crossing is at $x = (c_2/c_1)\mu_1$. It follows that $L(x)$ crosses c_1/μ_1 to the right of $(c_2/c_1)\mu_1$. Hence we should never plan replacement more frequently than $(c_2/c_1)\mu_1$ time units when F is IFR.

For the case in which x_0 uniquely satisfies (2.3) and minimizes $A(G)$, the resulting minimum attained is

$$A(x_0) = (c_1 - c_2)r(x_0) \tag{2.4}$$

as pointed out by Barlow and Hunter (1960c).

Infinite Time Span Solution Applied to Truncated Normal. A class of distributions of practical interest consists of the truncated normal distributions. An optimum age replacement interval for this class for an infinite time span is readily obtained using (2.3).

The density $f(x)$ of the truncated normal distribution may be written as

$$f(x) = \begin{cases} \dfrac{1}{b\sigma}\varphi\left(\dfrac{x-\mu}{\sigma}\right), & x \geq 0, \\ 0, & \text{otherwise,} \end{cases}$$

where
$$\varphi(x) = \frac{1}{\sqrt{2\pi}} e^{-x^2/2},$$

and
$$b = \frac{1}{\sigma} \int_0^\infty \varphi\left(\frac{x-\mu}{\sigma}\right) dx.$$

If $\mu/\sigma \geq 3$, then $f(x)$ is very close to the density of a normal distribution with mean μ and standard deviation σ.

Making the change of variable $y_0 = (x_0 - \mu)/\sigma$ and defining
$$r_N(x) = \frac{\varphi(x)}{\int_x^\infty \varphi(t)\, dt},$$
we have from (2.3)
$$K(y_0) = r_N(y_0) \int_{-\mu/\sigma}^{y_0} \int_t^\infty \frac{1}{\sqrt{2\pi}} e^{-x^2/2}\, dx\, dt$$
$$- \int_{-\mu/\sigma}^{y_0} \frac{1}{\sqrt{2\pi}} e^{-x^2/2}\, dx = \frac{bc_2}{c_1 - c_2}. \quad (2.5)$$

If $\mu/\sigma > 3$, we can approximate b by 1 and solve for $y_0 = (x_0 - \mu)/\sigma$ as a function of $c_2/(c_1 - c_2)$. Figure 2.2 is a graph of $K(y_0)$ for $\mu/\sigma = 3$.

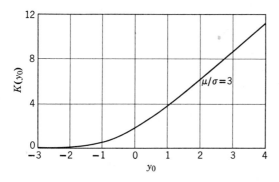

Fig. 2.2. $K(y_0)$ needed in obtaining strictly periodic policy for normal distribution. From Barlow and Proschan, Planned replacement, in *Studies in Applied Probability and Management Science*, edited by Arrow, Karlin, and Scarf, Stanford University Press, Stanford, Calif., 1962.

To obtain $A(x_0)$, the corresponding minimum average cost per unit time, we use Figure 2.3, which is a graph of $r_N(y_0)$, the failure rate function of a normal distribution with mean 0 and standard deviation 1. Since
$$r(x_0) = \frac{1}{\sigma} r_N(y_0),$$

we get from (2.4)

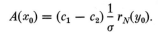

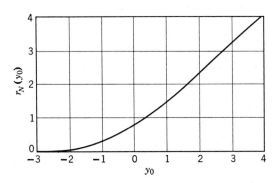

Fig. 2.3. Failure rate for the normal distribution $N(0, 1)$. From Barlow and Proschan Planned replacement, in *Studies in Applied Probability and Management Science* edited by Arrow, Karlin, and Scarf, Stanford University Press, Stanford, Calif., 1962.

Example 1. Many types of electron tubes used in commercial airline communication equipment and elsewhere tend to have a truncated normal failure distribution (Aeronautical Radio, Inc., 1958). A certain tube used in commercial equipment has a truncated normal failure distribution with a mean life of 9080 hours and a standard deviation σ of 3027. Suppose $c_1 = \$1100$ and $c_2 = \$100$, so that $c_1/c_2 = 11$. Setting $K(y_0) = c_2/(c_1 - c_2) = .1$, we see from Figure 2.2 that $y_0 = -1.5$, and hence $x_0 = 4540$ hours since $x_0 = \sigma y_0 + \mu$. From Figure 2.3 we find $r_N(y_0) = .14$, and

$$A(x_0) = \frac{1000}{3027}(.14) = \$.05.$$

Hence replacement costs us approximately 5 cents per hour on the average under the optimal replacement procedure. It costs us $c_1/\mu = \$.12$ per hour on the average if we simply replace failures as they occur. The optimal policy yields a 58 per cent reduction in cost.

Weiss (1956c) presents a somewhat more general version of age replacement for an infinite time span. His model differs from the previous model in that in addition to the effects of failure and replacement, the effects of repair (specifically, the time, considered random, and the cost of repair) are considered. Following the arguments just given, the asymptotic expected cost per unit of time may be obtained and minimized in some cases. Weiss presents several simplified examples of the calculation of optimal replacement policies but shows that an optimal policy need not always exist.

Minimax Strategy for an Infinite Time Span. In actual practice, we may not know the failure distribution F explicitly. Nevertheless, we can still obtain bounds on the expected number of failures, the expected number of planned replacements, etc., for various replacement policies under the assumption of an increasing failure rate, as shown in Chapter 3. However, to determine an *optimum* replacement policy we require rather explicit knowledge of F. In the absence of such explicit knowledge, an appropriate replacement strategy to seek might be the "minimax" strategy, that is, the replacement policy which minimizes the maximum expected cost, where the maximum is to be computed over all possible failure distributions.

THEOREM 2.2. *If we know only the mean μ_1 (and not F), the minimax age replacement policy for an infinite time span is to replace only at failure.*

Proof. Let $L(x, F)$ denote asymptotic expected cost per unit of time using replacement interval x when the failure distribution is F with mean μ_1. Let $F_0(t) = 1 - e^{-t/\mu_1}$. Then

$$\frac{c_1}{\mu_1} = \max_F L(\infty, F) \geq \min_x \max_F L(x, F) \geq \max_F \min_x L(x, F)$$

$$\geq \min_x L(x, F_0) = L(\infty, F_0) = \frac{c_1}{\mu_1}.$$

Hence the minimax strategy is never to schedule replacement. ∥

Note further that the corresponding failure distribution maximizing asymptotic expected cost per unit of time is the exponential.

THEOREM 2.3. *If we know only μ_1 and μ_2 (and not F) and*

$$\frac{\sigma}{\mu_1} \geq 1 - \frac{c_2}{c_1},$$

then the minimax age replacement strategy for an infinite time span is to replace only at failure.

Proof. Without loss of generality, assume $\mu_1 = 1$. Consider first the case where $\mu_2 \leq 2$. Let

$$F_0(t) = \begin{cases} 0, & 0 \leq t \leq 1 - \sqrt{\mu_2 - 1} \\ 1 - \exp\left(-\frac{t - 1 + \sqrt{\mu_2 - 1}}{\sqrt{\mu_2 - 1}}\right), & t > 1 - \sqrt{\mu_2 - 1}. \end{cases}$$

It is readily verified that F_0 has first moment μ_1, second moment μ_2. Note that Equation (2.3)

$$r_0(x) \int_0^x \bar{F}(t)\, dt - F_0(x) = \frac{c_2}{c_1 - c_2}$$

has no finite solution in x since the left side is 0 at $x = 0$, is $\leq c_2/(c_1 - c_2)$ at $x = \infty$, and is strictly increasing in between. This implies that the optimum replacement policy against F_0 is to replace only at failure.

Thus if $\mathscr{F}$ is the class of distributions F such that

$$\frac{\sigma}{\mu_1} \geq 1 - \frac{c_2}{c_1},$$

then $\quad \dfrac{c_1}{\mu_1} = \max_{F \in \mathscr{F}} L(\infty, F) \geq \min_{x} \max_{F \in \mathscr{F}} L(x, F) \geq \max_{F \in \mathscr{F}} \min_{x} L(x, F)$

$$\geq \min_{x} L(x, F_0) = L(\infty, F_0) = \frac{c_1}{\mu_1}.$$

Therefore the minimax strategy is to replace only at failure.

Now suppose $\mu_2 > 2$. Let $f_0(t) = \lambda^\alpha t^{\alpha-1} e^{-\lambda t}/\Gamma(\alpha)$ for $t \geq 0$, $F_0(t) = \int_0^t f_0(u)\, du$, where $\lambda = \alpha = 1/(\mu_2 - 1)$. F_0 is DFR so that the optimum replacement policy against F_0 is to replace only at failure. The rest of the argument is as before. ∥

It can be shown that for known F (IFR) for which

$$\frac{\sigma}{\mu_1} < 1 - \frac{c_2}{c_1}$$

the optimum replacement interval is finite.

Finite Time Span. The search for an optimum (nonrandom) age replacement policy is much more difficult when the time span is finite than when infinite. First we must show that such an optimum policy actually exists.

THEOREM 2.4. *Let the failure distribution F be continuous. Then there exists a minimum-cost age replacement policy for any finite time span $[0, t]$.*

Proof. For $\delta > 0$ sufficiently small the expected cost of using replacement interval $s < \delta$ will be greater than or equal to $c_1 M(t)$, where $M(t)$ is the renewal quantity associated with the distribution $F(t)$. We shall consider only replacement policies with replacement interval $s \geq \delta$. The interval $s \geq t$ is equivalent to no planned replacement.

Let $C_n(t, s)$ be the expected cost in $[0, t]$, considering only the first n removals following a policy of replacement at interval s. Clearly,

$$C_1(t, s) = \begin{cases} c_1 F(t), & t \leq s \\ c_1 F(s) + c_2 \bar{F}(s), & t > s. \end{cases}$$

For $n = 1, 2, \ldots$, we have

$$C_{n+1}(t, s) = \begin{cases} \int_0^t [c_1 + C_n(t - y, s)] \, dF(y), & t \leq s \\ \int_0^s [c_1 + C_n(t - y, s)] \, dF(y) + [c_2 + C_n(t - s, s)] \overline{F}(s), & t > s. \end{cases}$$

$C_1(t - js, s)$ is continuous in values of s for which t/s is not an integer for $j = 0, 1, 2, \ldots$, and hence so is $C_{n+1}(t - js, s)$. By induction, for t/s not an integer, $C_n(t, s)$ is continuous in s for $n = 1, 2, \ldots$.

Thus $\{C_n(t, s)\}$ is a sequence of functions continuous in s for t/s not an integer. But $C_n(t, s)$ converges to $C(t, s)$ uniformly for $s \geq \delta$, because (a) at most t/δ planned replacements can occur, so that for $n > t/\delta$ the probability of at least n removals caused by either failure or planned replacement is less than the probability of at least $n - [t/\delta]$ failures, and (b) this latter probability $\leq F^{(n-[t/\delta])}(t)$ where $[x]$ denotes the greatest integer contained in x, and (c) $\sum_{j=n}^{\infty} F^{(j-[t/s])}(t) \to 0$ as $n \to \infty$ uniformly in $s \geq \delta$. Hence $C(t, s)$ is continuous for s in $[t/(i + 1), t/i)$ for $i = 1, 2, \ldots$.

For any $\epsilon > 0$, consequently, $C(t, s)$ assumes its minimum for s in $[t/(i + 1), t/i - \epsilon]$ for $i = 1, 2, \ldots$. But

$$C(t, t/i) < \lim_{\epsilon \to 0^+} C(t, t/i - \epsilon),$$

since the two quantities differ essentially by the cost of one planned replacement. Thus $C(t, s)$ is continuous in s except at the point t/i, at which the function assumes a value lower than values immediately to the left for $i = 1, 2, \ldots$. Hence $C(t, s)$ assumes its minimum in s.

If this minimum is less than the cost of scheduling no planned replacements in $[0, t]$, we follow a policy of planned replacement with a replacement interval that minimizes expected costs. Otherwise, we schedule no planned replacements. Either way there exists a minimum-cost age replacement policy. ∥

Unfortunately, no general formulas are available for computing the optimum age replacement policy. We shall, however, consider a particular example to show how the optimal age replacement policy and the corresponding expected cost $C^*(t)$ might be calculated. Let $f(t) = te^{-t}$, a gamma density. Let the total cost of experiencing and replacing a failure be $c_1 = 10$ and of making a planned replacement be $c_2 = 1$. Find the replacement interval to use under an age replacement policy so as to minimize expected costs during $[0, t]$ for $0 < t \leq 6$.

From Theorem 2.4 we know that there exists a replacement interval s_t such that $C^*(t) = \min_s C(t, s) = C(t, s_t)$. Further, since

$$M(t) = \frac{t}{2} - \frac{1}{4} + \frac{1}{4}e^{-2t},$$

we have

$$C^*(t) = c_1 M(t) = 10\left(\frac{t}{2} - \frac{1}{4} + \frac{1}{4}e^{-2t}\right), \quad \text{if } s_t = \infty$$

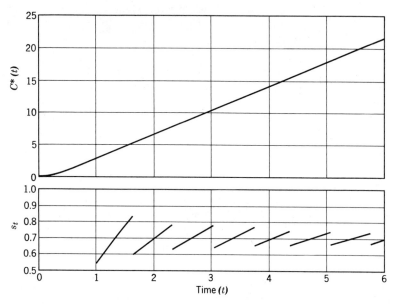

Fig. 2.4. Expected cost and replacement interval under optimal age replacement policy. From Barlow and Proschan, Planned replacement, in *Studies in Applied Probability and Management Science*, edited by Arrow, Karlin, and Scarf, Stanford University Press, Stanford, Calif., 1962.

(that is, no scheduled replacement) or

$$C^*(t) = \int_0^{s_t} [10 + C(t - y, s_t)]\, dF(y) + [1 + C(t - s_t, s_t)]\bar{F}(s_t)$$

if $0 < s_t < t$, where

$$\bar{F}(t) = 1 - F(t) = \int_t^\infty x e^{-x}\, dx = (1 + t)e^{-t}.$$

By numerical integration for values of $s = .01, .02, .03, \ldots, [t/.01].01$, we compute $C(t, s)$ from the integral equation

$$C(t, s) = \int_0^s [10 + C(t - y, s)]\, dF(y) + [1 + C(t - s, s)]\bar{F}(s).$$

We then take s_t as the value of s that minimizes $C(t, s)$. This has been done numerically for $0 \leq t \leq 6$. The values of $C^*(t)$, the expected cost under the optimal age replacement policy, and the corresponding optimal replacement interval s_t are plotted in Figure 2.4. Note that the solution may be obtained as before, even if the distribution function F is known only through an empirical estimate.

In Section 2.4 we shall compute the expected cost and the optimum sequential replacement policy under the same assumptions for costs and failure density. The two policies will then be compared.

2.2 Block replacement

Recall from Section 4 of Chapter 3 that under a policy of block replacement the unit is replaced at times kT ($k = 1, 2, \ldots$) *and* at failure.

Savage (1956) studies optimum block replacement policies for an infinite time span within a more general setting. His formulation does not seem readily applicable since he leaves the expression for the cost as a function of the replacement interval in general form.

Infinite Time Span. From Theorem 4.2 of Chapter 3 we deduce that the expected cost $B(T)$ per unit of time following a block replacement policy at interval T over an infinite time span is given by

$$B(T) = \frac{c_1 M(T) + c_2}{T} \qquad (2.6)$$

where, as usual, $M(T)$ is the renewal function (expected number of failures in $[0, T]$) corresponding to the underlying failure distribution F.

If $F(T)$ is continuous, it is clear that $B(T)$ is continuous for $0 < T < \infty$ with $B(T) \to \infty$ as $T \to 0$. If we interpret replacement at interval $T = \infty$ as replacement only at failure, it follows that $B(T)$ has a minimum for $0 < T \leq \infty$.

If F has a density f, it follows that $M(T)$ has a derivative $m(t)$, known as the renewal density [see Equation (2.4) of Chapter 3]. Hence a necessary condition that a finite value T_0 minimize $B(T)$ is that it satisfy (2.7), obtained by differentiating (2.6) and setting the derivative equal to 0:

$$Tm(T) - M(T) = \frac{c_2}{c_1}. \qquad (2.7)$$

The resulting value of $B(T_0)$ from (2.6) is $c_1 m(T_0)$.

Example. Let $f(t)$ be the gamma density of order 2:

$$f(t) = te^{-t}.$$

Then by inserting (2.13) and (2.14) of Chapter 3 in (2.7) and simplifying, we obtain

$$e^{-2T}(1 + 2T) = 1 - \frac{4c_2}{c_1}. \tag{2.8}$$

Note that for $c_2/c_1 \geq \frac{1}{4}$ no finite value of T satisfies (2.8), implying that replacement only at failure is optimal. For given $c_2/c_1 < \frac{1}{4}$, since the left side of (2.8) is strictly decreasing in T from an initial value of 1, a unique finite solution T_0 exists and may readily be found by using tables of the incomplete gamma function, such as in Karl Pearson (1934) or E. C. Molina (1942). The resulting minimum is

$$B(T_0) = c_1(\tfrac{1}{2} - \tfrac{1}{2}e^{-2T_0}).$$

2.3 Periodic replacement with minimal repair at failure

Barlow and Hunter (1960c) introduce the notion of periodic replacement or overhaul with minimal repair for any intervening failures. In this model it is assumed that the system failure rate remains undisturbed by any repair of failures between the periodic replacements. Barlow and Hunter show how to calculate the optimum period between replacements or overhauls for an infinite time span. In addition, they compare results obtained under this policy with those for the optimum age replacement policy, thus providing the decision maker with the information needed to choose between them.

The replacement policy of this section is discussed in Barlow and Hunter (1960c) under the designation "Policy II." We assume that after each failure only minimal repair is made so that system failure rate $r(t)$ (corresponding to failure distribution F having continuous density f) is not disturbed. Such a policy might apply to a complex system such as a computer, airplane, etc. Replacement or overhaul occurs at times $T, 2T, 3T, \ldots$. Our problem is to select T so as to minimize

$$C(T) = \lim_{t \to \infty} \frac{c_1 EN_1(t) + c_2 EN_2(t)}{t},$$

where $c_1 =$ cost of minimal repair,
$c_2 =$ cost of replacement (not necessarily $< c_1$),
$N_1(t) =$ number of failures in $[0, t]$,
$N_2(t) =$ number of replacements in $[0, t]$.

As we have seen in obtaining (2.6), $\dfrac{EN_2(t)}{t} \to \dfrac{1}{T}$ as $t \to \infty$. To evaluate $\lim\limits_{t \to \infty} \dfrac{EN_1(t)}{t}$, we first verify readily that the probability of at least one failure in an interval $(u, u + h)$ is $r(u)h + o(h)$, and the probability of two or more failures is $o(h)$. From this it follows that $EN_1(u + h) - EN_1(u)$, the expected number of failures in $(u, u + h)$, is $r(u)h + o(h)$. Since

$$EN_1(T) = [EN_1(h) - EN_1(0)]$$
$$+ [EN_1(2h) - EN_1(h)] + \cdots + [EN_1(T) - EN_1(kh)]$$

where $k = [T/h]$, it follows by the usual integral calculus arguments that $EN_1(T) = \int_0^T r(u)\,du$. Thus, using Theorem 4.2 of Chapter 3, we have

$$C(T) = \frac{c_1 \int_0^T r(u)\,du + c_2}{T}. \tag{2.9}$$

Noting that $-\log \bar{F}(T) = \int_0^T r(u)\,du$ and assuming F is IFR, we can obtain sharp bounds on $C(T)$ by using Tables 1, 2, and 3 of Appendix 3 when μ_1 or both μ_1 and μ_2 are specified.

To minimize $C(T)$ we set its derivative equal to 0, obtaining

$$\int_0^T [r(T) - r(u)]\,du = \frac{c_2}{c_1}. \tag{2.10}$$

THEOREM 2.5. *If there exists an interval $[0, b)$, where $0 < b \leq \infty$, such that $r(t)$ is continuous and unbounded on $[0, b)$, then (2.10) has a solution in T.*

Proof. Denote the left side of (2.10) by $L(T)$. Then $L(t)$ is continuous on $[0, b)$, with $L(0) = 0$. Let $\{T_i\}$ be a sequence such that $0 < T_1 < T_2 < \cdots$, $r(T) \leq r(T_i)$ for $T \leq T_i$, $i = 1, 2, \ldots$, and $\lim_{i \to \infty} r(T_i) = \infty$. Such a sequence exists since $r(t)$ is unbounded. Then $L(T_i) = T_i r(T_i) - \int_0^{T_i} r(u)\,du \geq T_1 r(T_i) - \int_0^{T_1} r(u)\,du \to \infty$ as $i \to \infty$. Thus $L(T)$ assumes all nonnegative values as T ranges from 0 to b. Hence (2.10) has a solution in T. ∥

A sufficient condition for a solution of (2.10) to be unique is given in

THEOREM 2.6. *Let $r(t)$ be strictly increasing and differentiable. Then if a solution to (2.10) exists, it is unique.*

Proof. Differentiating $L(T)$ yields $L'(T) = Tr'(T) > 0$ by hypothesis. Thus a solution to (2.10) must be unique. ∥

Note that if a unique solution to (2.10) exists, it must yield a minimum for $C(T)$, since $C(T) \to \infty$ as $T \to 0$.

If T_0 is the value of T minimizing $C(T)$, then from (2.10) the resulting minimum value of $C(T)$ is
$$C(T_0) = c_1 r(T_0).$$

Example. Suppose F is a Weibull distribution function,
$$F(t) = 1 - e^{-\lambda t^\alpha}$$
where $\lambda > 0$, $\alpha > 1$. Then $r(t) = \lambda \alpha t^{\alpha-1}$ and $\mu = \Gamma\left(\frac{1}{\alpha} + 1\right) / \lambda^{1/\alpha}$. By using (2.10) the optimum T_0 satisfies
$$\lambda(\alpha - 1)T_0^\alpha = \frac{c_2}{c_1},$$
so that
$$T_0 = \left[\frac{c_2}{\lambda(\alpha - 1)c_1}\right]^{1/\alpha}.$$

2.4 Sequential replacement over a finite time span

The notion of sequential replacement is relevant only for a finite time span. Under a sequential policy the age for which replacement is scheduled is no longer the same following successive removals but depends on the time still remaining. Clearly the added flexibility permits the achievement of an optimum sequential policy having lower cost than that of the corresponding optimum age replacement policy. Barlow and Proschan (1962) prove the existence of an optimum sequential policy, obtain qualitative results concerning its form, and illustrate its calculation for the same example for which the optimum age replacement policy was calculated; for this case the two policies are directly compared.

Under sequential replacement, the *next* planned replacement interval after each removal is selected to minimize expected expenditure during the remaining time. Thus we do not specify at the beginning of the original time span each future planned replacement interval; rather, after each removal we specify only the next replacement interval. This gain in flexibility leads to reduction in expected cost. The following analysis is based on Barlow and Proschan (1962).

Our first step will be to show that within the class of sequential policies we need only consider nonrandom replacement. To designate the policy under which distribution G_τ governs the next replacement interval when τ units of time are left following a removal, $0 \leq \tau \leq t$, t being the initial time span, we write $\{G_\tau: 0 \leq \tau \leq t\}$. By $\{D_t;\ G_\tau: 0 \leq \tau \leq t\}$ we mean that distribution D_t governs the replacement interval when t units of time remain, and distribution G_τ governs the replacement interval when τ units of time remain, $0 \leq \tau < t$. Note that with probability $\bar{G}_\tau(\tau)$ no planned replacement is scheduled in the remaining time span τ.

OPTIMUM MAINTENANCE POLICIES 99

LEMMA. Given any policy $P_1 = \{G_\tau: 0 \leq \tau \leq t\}$, there exists a policy $P_2 = \{D_t; G_\tau: 0 \leq \tau < t; D_t$ a degenerate distribution$\}$ with no greater expected cost.

Proof. Let $S_i(t)$ denote expected cost during $[0, t]$ under P_i, with $i = 1, 2$. If $S_1(t) = \infty$, the conclusion is obvious.

Assume then that $S_1(t) < \infty$. We may write $S_1(t) = \int_0^\infty R(x) \, dG_t(x)$, where

$$R(x) = \begin{cases} \int_0^x [c_1 + S_1(t-y)] \, dF(y) + [c_2 + S_1(t-x)] \bar{F}(x), & 0 \leq x \leq t, \\ \int_0^t [c_1 + S_1(t-y)] \, dF(y), & x > t. \end{cases}$$

Define r as $\inf_{0 \leq x \leq t} R(x)$. If $R(x_0) \leq r$ for any $x_0 > t$, choose D_t assigning unit mass at x_0 (that is, schedule no planned replacement initially). Then $S_2(t) \leq S_1(t)$.

If $R(x) > r$ for $x > t$, and if (1) r is actually attained, say at x_0 ($0 \leq x_0 \leq t$), choose D_t assigning unit mass at x_0; if (2) r is not attained in $[0, t]$, then $R(x) > r$ for $0 \leq x \leq t$. Thus $S_1(t) > r$. Specifically, let $S_1(t) = r + \delta$, where $\delta > 0$. Since $r = \inf_{0 \leq x \leq t} R(x)$, there exists x_1 in $[0, t]$ such that $R(x_1) < r + \delta$. Choose D_t assigning unit mass at x_1. Then $S_2(t) < r + \delta$. In either case (1) or case (2), $S_2(t) \leq S_1(t)$. ∥

By repeated applications of the lemma, we immediately obtain the following theorem.

THEOREM 2.7. Given any policy $P_1 = \{G_\tau: 0 \leq \tau \leq t\}$, there exists a policy $P_2 = \{D_\tau$, a degenerate distribution: $0 \leq \tau \leq t\}$ with no greater expected cost.

That is, for any sequential policy calling for random replacement intervals, there exists a sequential policy requiring only nonrandom replacement intervals with no greater expected cost. Thus we need not consider random replacement policies in searching for the optimal sequential policy.

Define $L(t)$ as the infimum of expected cost during $[0, t]$ over all sequential policies. We use "infimum" rather than "minimum," since we do not yet know that an optimal policy, that is, one achieving $L(t)$, exists. We present certain basic properties of $L(t)$ useful in establishing the existence of an optimal policy.

THEOREM 2.8. $L(t)$ is nondecreasing.

Proof. Given $\delta > 0$ and $\epsilon > 0$, there exists a policy $\{G_\tau: 0 \leq \tau \leq t + \delta\}$ with expected cost $S(t + \delta)$ during $[0, t + \delta]$ less than $L(t + \delta) + \epsilon$. Define $\hat{G}_\tau(x) = G_{\tau+\delta}(x)$ for $0 \leq \tau \leq t$ and $0 \leq t < \infty$. Then if $\hat{S}(t)$ is the expected cost during $[0, t]$ of the policy $\{\hat{G}_\tau: 0 \leq \tau \leq t\}$, we must have

$$L(t + \delta) + \epsilon > \hat{S}(t + \delta) \geq \hat{S}(t) \geq L(t).$$

Since ϵ is arbitrary, $L(t + \delta) \geq L(t)$. ‖

Note, however, that it is *not* true that for an arbitrary policy for $[0, t_0]$ the expected cost $S(t)$ during $[0, t]$ is a nondecreasing function of t ($0 \leq t \leq t_0$). As a simple counterexample, let $0 < t_1 < t_2$ and let F be absolutely continuous. Consider a policy in which G_{t_1} concentrates at $t_1/2$ and G_{t_2} concentrates at ∞ (that is, no planned replacement). Then the expected cost $S(t_1) \geq c_2$, whereas $S(t_2)$ can be made less than c_2 for sufficiently small t_2.

Next we shall obtain the useful result.

THEOREM 2.9. Let F be continuous. Then $L(t)$ is continuous.

Proof. Given $\epsilon > 0$, there exists a policy $\{G_\tau: 0 \leq \tau \leq t\}$ with expected cost $S(t)$ during $[0, t]$ less than $L(t) + \epsilon/2$. Define $\hat{G}_{\tau+\delta}(x) = G_\tau(x)$ for $0 < \tau \leq t$ and $0 \leq x < \infty$, whereas for $0 \leq \tau \leq \delta$, $\hat{G}_\tau$ calls for no planned replacement. Let $\hat{S}(t + \delta)$ represent the expected cost of the policy $\{\hat{G}_\tau: 0 \leq \tau \leq t + \delta\}$. Then

$$L(t + \delta) - L(t) < \hat{S}(t + \delta) - S(t) + \frac{\epsilon}{2}$$

$$\leq \frac{\epsilon}{2} + c_1 \sup_{0 \leq \tau \leq t} [M(\tau + \delta) - M(\tau)],$$

where M is the renewal quantity corresponding to F. Since F is continuous, M is continuous and thus is uniformly continuous on any closed interval. The conclusion that $L(t)$ is continuous readily follows. ‖

Having established the continuity of $L(t)$, we may now show the existence of an optimal policy, that is, one attaining $L(t)$. First we define

$$W(x, t) = \int_0^x [c_1 + L(t - y)] \, dF(y) + [c_2 + L(t - x)] \bar{F}(x).$$

Note that $W(x, t)$ represents the expected cost if initial replacement is scheduled for time $x \leq t$ and if the infimum of expected cost during the remaining time is then attained.

LEMMA. Let F be continuous. Then $W(x, t)$ attains a minimum as x ranges over $[0, t]$.

Proof. $\int_0^x [c_1 + L(t - y)] \, dF(y)$ is continuous in x because the integrand is bounded by Theorem 2.8. Furthermore, $L(t - x)$ is continuous in x from Theorem 2.9. Hence $W(x, t)$ is continuous in x over $[0, t]$ and so attains its minimum over the interval. ‖

Next we define quantities needed to actually construct the optimal policy. Let

$$U(t) = \min_{0 \leq x \leq t} W(x, t),$$

$$V(t) = \int_0^t [c_1 + L(t - y)] \, dF(y),$$

$$x_t = \begin{cases} \infty, & \text{if } U(t) > V(t) \\ x, & \text{for which } W(x, t) = U(t) \text{ if } U(t) \leq V(t). \end{cases}$$

We may interpret $U(t)$ as the infimum of the expected cost if initial replacement is scheduled during $[0, t]$, $V(t)$ as the infimum of the expected cost if no initial replacement is scheduled, and x_t as the best time to schedule initial replacement when t time units remain (with $x_t = \infty$ indicating no initial scheduled replacement). We shall use $\{u_\tau : 0 \leq \tau \leq t\}$ to refer to the policy in which u_τ is the next replacement interval when τ units of time remain.

THEOREM 2.10. *Let F be continuous. Then the policy $\{x_\tau : 0 \leq \tau \leq t\}$ has expected cost $L(t)$ and so is optimal.*

Proof. By Theorem 2.7 we may confine ourselves to nonrandom replacement policies. Let $S(t)$ refer to the expected cost of a typical policy. Then

$$L(t) = \min \left\{ \inf_{\substack{0 \leq x \leq t \\ \text{all policies}}} \left[\int_0^x [c_1 + S(t - y)] \, dF(y) + [c_2 + S(t - x)] \bar{F}(x) \right], \right.$$

$$\left. \inf_{\text{all policies}} \left[\int_0^t [c_1 + S(t - y)] \, dF(y) \right] \right\}$$

$$\geq \min \left\{ \inf_{0 \leq x \leq t} \left[\int_0^x [c_1 + L(t - y)] \, dF(y) + [c_2 + L(t - x)] \bar{F}(x) \right], \right.$$

$$\left. \int_0^t [c_1 + L(t - y)] \, dF(y) \right\}$$

$$= \min \left\{ \min_{0 \leq x \leq t} \left[\int_0^x [c_1 + L(t - y)] \, dF(y) + [c_2 + L(t - x)] \bar{F}(x) \right], \right.$$

$$\left. \int_0^t [c_1 + L(t - y)] \, dF(y) \right\},$$

since by the lemma to Theorem 2.10 the infimum is actually attained. Thus

$$L(t) \geq \min \left[E(L_1; x_t \leq t) + \int_0^{x_t} L(t - y) \, dF(y) + L(t - x_\tau) \bar{F}(x_\tau), \right.$$

$$\left. E(L_1; x_t = \infty) + \int_0^t L(t - y) \, dF(y) \right],$$

where L_i equals the expected cost of the ith removal (considered 0 if fewer than i removals occur in the remaining time span).

By repeated applications of this argument, we obtain

$$L(t) \geq \sum_{i=1}^{n} EL_i + R(n), \quad n = 1, 2, \ldots,$$

where $R(n) \geq 0$ is the remainder at the nth stage. But $\sum_{i=1}^{n} EL_i$ is an increasing function of n bounded by $L(t)$; hence $\sum_{i=1}^{n} EL_i$ approaches a limit $A \leq L(t)$. But $A \geq L(t)$, because $L(t)$ is the infimum of expected cost over all policies; hence $A = L(t)$. Thus the policy $\{x_\tau : 0 \leq \tau \leq t\}$ has expected cost $L(t)$ and is therefore optimal. ‖

Note we may write $L(t) = \min [U(t), V(t)]$.

Next we shall obtain the important result that, for failure distributions having a nondecreasing failure rate, under an optimal policy there exists t_0 such that planned replacement is required when the remaining time is greater than t_0, whereas planned replacement is not called for when the remaining time is less than t_0. (Note that t_0 may be infinite, as in the case of the exponential failure density, so that scheduled replacement is never required.) This result simplifies the analysis considerably and permits a simplified procedure for computation.

LEMMA. Let $L_i(t)$ be the loss under an optimal policy when the initial component has continuous failure distribution F_i for $i = 1, 2$, and $F_1(x) \leq F_2(x)$ for $x \geq 0$, and subsequent replacements have continuous failure distribution F. Then $L_1(t) \leq L_2(t)$ for $t \geq 0$.

Proof. Write

$$U_i(t) = \min_{0 \leq x \leq t} \left[\int_0^x [c_1 + L(t-y)] \, dF_i(y) + [c_2 + L(t-x)] \bar{F}_i(x) \right],$$

where L refers to the expected loss that results from following an optimal policy under failure distribution F. Integrating by parts, we obtain

$$U_i(t) = \min_{0 \leq x \leq t} \Big[[c_1 + L(t-x)]F_i(x) \\ - \int_0^x F_i(y) \, dL(t-y) + [c_2 + L(t-x)]\bar{F}_i(x) \Big].$$

Thus

$$U_2(t) - U_1(t) \geq (c_1 - c_2)[F_2(x) - F_1(x)] \\ - \int_0^x [F_2(y) - F_1(y)] \, dL(t-y),$$

OPTIMUM MAINTENANCE POLICIES

where x is the minimizing value yielding $U_2(t)$. But $F_2 - F_1 \geq 0$ by hypothesis, and $L(t - y)$ is decreasing in y by Theorem 2.8. Hence $U_2(t) \geq U_1(t)$.

Similarly,

$$V_2(t) - V_1(t) = c_1[F_2(t) - F_1(t)] + \int_0^t [F_2(y) - F_1(y)] \, dL(t - y) \geq 0.$$

Hence $V_2(t) \geq V_1(t)$. Thus

$$L_2(t) = \min \, [U_2(t), V_2(t)] \geq [U_1(t), V_1(t)] = L_1(t). \; \|$$

THEOREM 2.11. Let F be an absolutely continuous failure distribution having an increasing failure rate. Then under an optimal policy there exists a value t_0 (possibly infinite) such that planned replacement is required when the remaining time $t > t_0$ and not required when the remaining time $t < t_0$.

Proof. If $x_t = \infty$ for all t, the theorem is trivially true. Assume then that there is a value τ for which $x_\tau < \tau$.

Consider a remaining time of $T > \tau$. We shall show that the expected cost $S_1(t)$ of following a policy of scheduling replacement after an elapsed time of $T - \tau + x_\tau$ time units (policy 1) is no greater than $S_2(t)$, the expected cost of scheduling no initial replacement (policy 2). (We assume the optimal policy is followed for all subsequent replacements under either policy.) This will imply that $x_T < T$.

The only difference in cost under the two policies will occur if the initial component survives until $T - \tau + x_\tau$ time units have elapsed, that is, until $\tau - x_\tau$ time units remain. Thus $S_1(T) \leq S_2(T)$ if

$$c_2 + L(\tau - x_\tau) \leq L(\tau - x_\tau \mid F_{T-\tau+x_\tau}),$$

where $L(\tau - x_\tau \mid F_{T-\tau+x_\tau})$ is the expected cost when $\tau - x_\tau$ time units remain and the component currently in operation is $T - \tau + x_\tau$ old. But we know that $c_2 + L(\tau - x_\tau) \leq L(\tau - x_\tau \mid F_{x_\tau})$, since the optimal policy calls for planned replacement at time x_τ when the time span is τ. Moreover, by the lemma to Theorem 2.11 and the assumption that F has an increasing failure rate, we know that

$$L(\tau - x_\tau \mid F_{T-\tau+x_\tau}) \geq L(\tau - x_\tau \mid F_{x_\tau}).$$

Thus $$L(\tau - x_\tau \mid F_{T-\tau+x_\tau}) \geq c_2 + L(\tau - x_\tau),$$

implying that $S_1(T) \leq S_2(T)$, which in turn implies that $x_T < T$. $\|$

It is worthwhile to note that the requirement in Theorem 2.11 that F have an increasing failure rate cannot be completely dropped. Consider the following example. Although we use a discrete failure distribution, we can approximate it arbitrarily closely with an absolutely continuous distribution without changing the conclusion.

Let failure occur at time 1 with probability p and at time 3 with probability $1 - p$. Then for the time interval $[0, 1 + \epsilon]$ (ϵ small), planned replacement just before 1 unit of time has elapsed costs c_2, whereas no planned replacement costs pc_1. Assume that $c_2 < pc_1$. Thus planned replacement just before 1 unit of time has elapsed is the optimal policy for the interval $[0, 1 + \epsilon]$. Next consider the time interval $[0, 2 + \epsilon]$. If initially planned replacement is not scheduled, the expected cost is $p(c_1 + c_2)$, since the probability of failure at time 1 is p, and failure at time 1 is followed by a planned replacement at time 2 under the optimal policy. If replacement is scheduled just before time 1, the expected cost is $2c_2$; if it is scheduled at time 2, the expected cost is $p(c_1 + c_2) + (1 - p)c_2$. If we pick p, c_1, and c_2 such that $p(c_1 + c_2) < 2c_2$, then the optimal policy for $[0, 2 + \epsilon]$ calls for no initial planned replacement. [For example, let $c_1 = .9$, $c_2 = .1$, and pick p in the interval $(\frac{1}{9}, \frac{2}{10})$.]

Next we shall describe a procedure for computing the optimal policy and its expected cost as a function of time. An illustrative example is worked to facilitate exposition. The computational procedure is quite direct and rapid, requiring only four minutes on the IBM 704 for the example.

As an illustration, we shall determine the optimal sequential policy and its expected cost when $f(t) = te^{-t}$, $c_1 = 10$, and $c_2 = 1$. This is the same example for which we computed the optimal age replacement policy. We shall perform our computations at successive points in time spaced at a small common interval of .01. First we compute $L(.01) = V(.01) = c_1 f(.01).01$; this represents the expected cost when the horizon time is .01, since we have taken the increments of time small enough to preclude scheduled replacement when only .01 units of time are left (that is, $x_{.01} = \infty$). Next we compute

$$V(.02) = \int_0^{.02} [c_1 + L(t - y)] \, dF(y)$$

which is approximated by

$$[c_1 + L(.01)]f(.01).01 + c_1 f(.02).01.$$

We also compute

$$W(.01, .02) = \int_0^{.01} [c_1 + L(.02 - y)] \, dF(y) + [c_2 + L(.02 - .01)]\bar{F}(.01)$$

$$\approx [c_1 + L(.01)]f(.01).01 + [c_2 + L(.01)]\bar{F}(.01).$$

(More generally, we would compute $W(x, t)$ for each increment value of $x < t$; in the present case there happens to be only one nontrivial value.)

It turns out that $V(.02) \leq W(.01, .02)$, so that scheduled replacement is not required when the horizon time is .02; thus $L(.02) = V(.02)$. In a similar fashion we compute $V(t)$ and $W(x, t)$ for $x < t$ by numerical integration for successive time points spaced at common interval .01. For each value of t we determine $U(t) = \min_{0<x<t} W(x, t)$ and compare it with $V(t)$. As long as $U(t)$ remains greater than $V(t)$ we continue to set $L(t) = V(t)$ and require no scheduled replacement. However, because for $t = .96$ we find that $U(.96)$ becomes less than or equal to $V(.96)$, we schedule replacement at time $x_{.96} = .50$, where $x_{.96} = .50$ yields $\min_{0<x<.96} W(x, .96)$. Furthermore, for any $t > .96$ we no longer need to compute $V(t)$ by Theorem 2.11, since

$$\frac{f(t)}{\bar{F}(t)} = \frac{te^{-t}}{e^{-t}(1+t)} = \frac{t}{1+t}$$

is an increasing failure rate.

Thus for $t > .96$ we compute only $W(x, t)$ for $0 < x < t$, and set $L(t) = \min_{0<x<t} W(x, t)$. The value x_t minimizing $W(x, t)$ is the scheduled replacement interval. Figure 2.5 shows $L(t)$ and x_t as functions of time for $0 \leq t < 6$; for $0 \leq t < .96$, $x_t = \infty$ and is not shown. It is interesting to note that x_t is discontinuous and piecewise monotonic increasing. Note further that the limiting value of x_t as $t \to \infty$ is given in (2.3) as the solution to

$$\frac{x}{1+x} \int_0^x e^{-t}(1+t)\, dt - [1 - e^{-x}(1+x)] = .11,$$

or $x_t = .68$.

If $f(0) \neq 0$, one additional complication enters. Note that in order to compute $W(x, t)$, $U(t)$, and $V(t)$, we must use the as yet unknown $L(t)$. Two methods are available for overcoming this difficulty.

1. We can approximate $L(t)$ by extrapolation. If the resulting value of $L(t)$ computed from the recursion differs, we can insert the new value and recompute $L(t)$.

2. If we are in a domain of values of t for which $L(t) = U(t)$, we transpose so that the unknown $L(t)$ appears on the left-hand side only. We can then solve for $L(t)$ (keeping in mind that $L(t)$ and $U(t)$ are the same) and compute the result. We solve similarly if $L(t) = V(t)$.

In Figure 2.6 we give a graphical comparison of expected cost under the optimal sequential policy and the optimal strictly periodic policy. Note that the difference remains quite small, and for $t = 6$ represents only about one per cent of the expected cost. In Figure 2.7 we make a

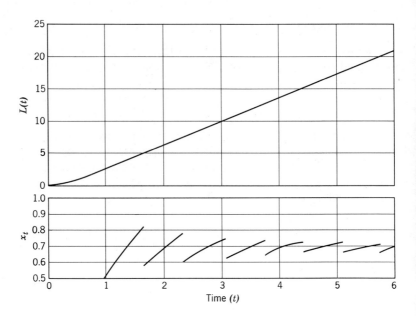

Fig. 2.5. Expected cost and replacement interval under optimal sequential policy. From Barlow and Proschan, Planned replacement, in *Studies in Applied Probability and Management Science,* edited by Arrow, Karlin, and Scarf, Stanford University Press, Stanford, Calif., 1962.

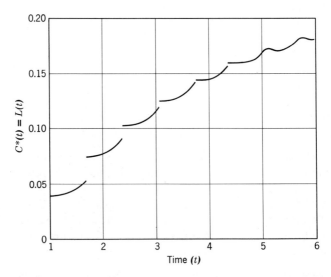

Fig. 2.6. Comparison of expected costs under optimal sequential and age replacement policies. From Barlow and Proschan, Planned replacement, in *Studies in Applied Probability and Management Science,* edited by Arrow, Karlin, and Scarf, Stanford University Press, Stanford, Calif., 1962.

OPTIMUM MAINTENANCE POLICIES

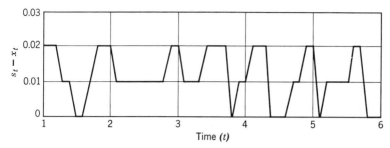

Fig. 2.7. Comparison of replacement intervals under optimal sequential and age replacement policies. From Barlow and Proschan, Planned replacement, in *Studies in Applied Probability and Management Science*, edited by Arrow, Karlin, and Scarf, Stanford University Press, Stanford, Calif., 1962.

similar graphical comparison of the optimal replacement interval under the two types of policies. Again the differences are small.

3. INSPECTION POLICIES

In Section 2 we discussed the problem of finding the best member within various classes of replacement policies. For each policy we assumed that detection and replacement of a failed unit occurred instantly. In the present section we assume that failure is discovered only by actual inspection and, in general, only after some time has elapsed since the occurrence of the failure. Our goal will be to determine inspection policies (i.e., schedules of inspection times) which minimize the total expected cost resulting from both inspection and failure.

The inspection models considered in this section arise in connection with systems which are deteriorating. Deterioration is assumed stochastic, and the condition of the system is known only through inspection. Optimization consists of minimizing total costs arising from inspection and undetected failure. Barlow, Hunter, and Proschan (1963) show how to obtain optimum inspection schedules for a broad class of failure distributions. A related model, the so-called "preparedness" model, is considered by Savage (1956) in a general context. Derman (1961) obtains the minimax inspection schedule in the absence of any information concerning the failure distribution; in his model an inspection may not detect a failure even when present, and the time span is finite. A new aspect of preparedness models is introduced in Jorgenson and Radner (1962) and generalized in Radner and Jorgenson (1962); they prove the optimality under certain conditions of "opportunistic" maintenance in which the maintenance action to be taken on a given part depends on the state of the rest of the system.

3.1 Minimizing expected cost until detection of failure

In our first model we assume that upon detection of failure the problem ends; specifically, no replacement or repair takes place. As an example, consider the problem of detecting the occurrence of an event (say, the arrival of an enemy missile or the presence of some grave illness such as cancer) when the time of occurrence is not known in advance. Each inspection involves a cost so that we do not wish to check too often. On the other hand, there is a penalty cost associated with the lapsed time between occurrence and its detection so that we wish to check often enough to avoid a long lapse of time between failure and detection.

Specifically, assume that (*a*) system failure is known only through checking, (*b*) checking takes negligible time and does not degrade the system, (*c*) the system cannot fail while being checked, (*d*) each check entails a fixed cost c_1, (*e*) the time elapsed between system failure and its discovery at the next check costs c_2 per unit of time, and (*f*) checking ceases upon discovery of failure. This model is treated by Barlow, Hunter, and Proschan (1963); some of the material in this section is taken from that paper.

If we let $N(t)$ denote the number of checks during $[0, t]$ and γ_t the interval between failure and discovery if failure occurs at time t, the corresponding cost is $c_1[N(t) + 1] + c_2\gamma_t$. (We are assuming a check always follows failure.) If the time of failure is governed by failure distribution F, the expected cost C is given by

$$C = \int_0^\infty \{c_1[EN(t) + 1] + c_2 E\gamma_t\} \, dF(t). \tag{3.1}$$

An optimum checking procedure is a specification of successive check times $x_1 < x_2 < x_3 < \cdots$ (possibly random) for which C is minimized.

As for the sequential replacement model of Section 2.4, it may be shown that any random schedule may be improved upon by a nonrandom schedule. Thus we shall confine attention to inspection schedules in which the times of inspection $x_1 < x_2 < x_3 < \cdots$ are constants. We shall define $x_0 = 0$ and require that the sequence $\{x_n\}$ be such that the support of F is contained in $[0, \lim_{n \to \infty} x_n)$. The latter assumption precludes the unacceptable possibility of undetected failure occurring with positive probability. From (3.1) the expected cost is explicitly

$$C = \sum_{k=0}^\infty \int_{x_k}^{x_{k+1}} [c_1(k + 1) + c_2(x_{k+1} - t)] \, dF(t). \tag{3.2}$$

As proved in Barlow, Hunter, and Proschan (1963), if F is continuous with finite mean, an optimum checking schedule must exist.

When it is known that failure must occur by time $T < \infty$, it is possible to give a necessary and sufficient condition on the failure distribution F to ensure that we need only check at the end of the interval.

THEOREM 3.1. Let $F(t) = 1$ for $t \geq T$. If

$$F(t) \leq \frac{1}{1 + (c_2/c_1)(T - t)}$$

for $0 \leq t \leq T$, the optimum checking policy will consist of a single check performed at time T. Conversely, if

$$F(t) > \frac{1}{1 + (c_2/c_1)(T - t)}$$

for some $0 \leq t \leq T$, the optimum checking policy will require, in addition to the check at time T, one or more checks before time T.

Proof. If a single check is performed at time T, the expected cost is

$$c_1 + c_2 \int_0^T (T - t) \, dF(t) = c_1 + c_2 \int_0^T F(t) \, dt.$$

If an additional check is performed at time x, the expected cost is

$$\int_0^x [c_1 + c_2(x - t)] \, dF(t) + \int_x^T [2c_1 + c_2(T - t)] \, dF(t)$$

$$= c_1 + c_1[1 - F(x)] + c_2 \int_0^T F(t) \, dt - c_2(T - x)F(x).$$

Thus a single check at time T is preferable if

$$c_1[1 - F(x)] - c_2(T - x)F(x) \geq 0, \qquad 0 \leq x \leq T$$

or
$$F(x) \leq \frac{1}{1 + (c_2/c_1)(T - x)}, \qquad 0 \leq x \leq T. \tag{3.3}$$

Moreover, this implies that a single check at time T is preferable to a single check at time T preceded by *two* checks at, say, times x_1 and x_2. Given that a check has occurred at time x_1, a single check before T, namely at time x_2, is warranted if and only if

$$\frac{F(x_2) - F(x_1)}{\overline{F}(x_1)} > \frac{1}{1 + (c_2/c_1)[(T - x_1) - (x_2 - x_1)]}. \tag{3.4}$$

But (3.4) implies

$$F(x_2) > \frac{1}{1 + (c_2/c_1)(T - x_2)}$$

because
$$F(x_2) > \frac{F(x_2) - F(x_1)}{\bar{F}(x_1)}.$$
This contradicts our assumption that
$$F(x) \leq \frac{1}{1 + (c_2/c_1)(T-x)}, \quad 0 \leq x \leq T.$$
Similarly, it is easy to see that (3.3) implies that a single check at time T is preferable to n checks before time T.

Conversely, if for some x in $[0, T]$ (3.3) is not true, checks at times x and T yield lower expected cost than does a single check at time T. Thus the optimum checking policy will require, in addition to the check at time T, one or more checks before time T. ‖

An interesting application of Theorem 3.1 will occur in constructing Example 1.

If the failure density f is a Pólya frequency function of order 2(PF_2) (see Appendix 1), the derivation of the optimum checking schedule becomes quite tractable. We will need some of the properties of PF_2 densities presented in Appendix 1.

Assuming a failure density f, a necessary condition that a sequence $\{x_k\}$ be a minimum cost checking procedure is that $\partial C/\partial x_k = 0$ for all k. Hence using (3.2) we obtain for $k = 1, 2, 3, \ldots$,

$$x_{k+1} - x_k = \frac{F(x_k) - F(x_{k-1})}{f(x_k)} - \frac{c_1}{c_2}. \tag{3.5}$$

When $f(x_k) = 0$, $x_{k+1} - x_k = \infty$ so that no more checks are scheduled. The sequence is determined recursively once we choose x_1.

THEOREM 3.2. *If the failure density f is PF_2, and $f(x) > 0$ for $x > 0$, the optimum checking intervals are nonincreasing.*

Proof. Let $\delta_k = x_{k+1} - x_k$, $\{x_k^*\}$ denote the optimum checking policy, and $\delta_k^* = x_{k+1}^* - x_k^*$, where $\{x_k\}$, $\{x_k^*\}$ satisfy (3.5). Assume for some k,

$$\frac{\delta_{k-1}}{\delta_{k-2}} = r > 1.$$

Then we shall show that $\delta_k/\delta_{k-1} > 1$; moreover, if $x_{k-2} \geq m$ where m is the mode of f, then $\delta_k/\delta_{k-1} \geq r$. (By Theorem 2 of Appendix 1, f must be unimodal.)

Note that

$$\delta_k - \delta_{k-1} = \frac{F(x_k) - F(x_{k-1})}{f(x_k)} - \frac{F(x_{k-1}) - F(x_{k-2})}{f(x_{k-1})}.$$

OPTIMUM MAINTENANCE POLICIES   111

By Theorem 1 of Appendix 1, the right-hand side is nonnegative since $\delta_{k-1} \geq \delta_{k-2}$ by assumption. Thus $\delta_{k-1} \geq \delta_{k-2}$ implies $\delta_k \geq \delta_{k-1}$.

If now $x_{k-2} \geq m$, then

$$\delta_k - r\delta_{k-1} = \frac{F(x_k) - F(x_{k-1})}{f(x_k)} - r\frac{F(x_{k-1}) - F(x_{k-2})}{f(x_{k-1})} + \frac{c_1}{c_2}(r-1)$$

is nonnegative by Theorem 3 of Appendix 1. Hence $\delta_{k-1} = r\delta_{k-2}$ and $x_{k-2} \geq m$ imply $\delta_k \geq r\delta_{k-1}$. We may thus conclude that if for any k, $\delta_k > \delta_{k-1}$, then $\delta_n \to +\infty$ geometrically fast from some point on as $n \to \infty$.

We will show that $\delta_k^* > \delta_{k-1}^*$ for some k contradicts the fact that $\{x_k^*\}$ is an optimum checking policy. Note that x_{n+1}^* is the optimum first checking point for the conditional density $f(t + x_n^*)/\bar{F}(x_n^*)$. Let $\alpha(x_n^*)$ denote the mean of this conditional density. Since $f(t + x_n^*)/\bar{F}(x_n^*)$ is PF_2,

$$\alpha(x_n^*) \leq \frac{\bar{F}(x_n^*)}{f(x_n^*)}$$

by Equation (7.1) of Chapter 2. Thus

$$\lim_{n \to \infty} \alpha(x_n^*) \leq \lim_{n \to \infty} \frac{\bar{F}(x_n^*)}{f(x_n^*)} < \infty$$

by Theorem 1 of Appendix 1. Hence the expected cost for the optimum first check using the density $f(t + x_n^*)/\bar{F}(x_n^*)$ is bounded above uniformly in n. But this cost is

$$c_1 + c_2 \int_{x_n^*}^{x_{n+1}^*} (x_{n+1}^* - t) \frac{f(t)\,dt}{\bar{F}(x_n^*)} = c_1 + c_2 \int_0^{\delta_n} (\delta_n - t) \frac{f(t + x_n^*)\,dt}{\bar{F}(x_n^*)} \to \infty$$

since $\delta_n \to +\infty$. This leads to the desired contradiction. Hence the optimum checking intervals are nonincreasing. ∥

The following example shows that the optimum checking times are not in general more and more frequent.

Example 1. Suppose an item fails according to a distribution which places mass p at time 1 and mass $1 - p$ at time 3. Alternatively, we can imagine a density function approaching this two-spiked density. Theorem 3.1 implies that it pays to check at times 1 and 3 if

$$p > \frac{c_1}{c_1 + 2c_2},$$

and in this case the checking intervals increase rather than decrease. The density is not PF_2, however, since it is bimodal (see Theorem 2 of Appendix 1).

Calculating Optimum Sequential Procedures. Theorem 3.3 gives us a powerful tool for computing optimum checking procedures when f is PF_2. To obtain Theorem 3.3 we need the following lemma:

LEMMA. Assume f is PF_2. Let $\{x_k\}$ be generated by (3.5) with $\delta_k > 0$ for $k = 1, 2, \ldots, N$, where $\delta_k = x_{k+1} - x_k$. Then

$$\frac{d\delta_k}{dx_1} \geq \frac{f(x_1)}{f(x_k)} \frac{d\delta_1}{dx_1}$$

for $k = 1, 2, \ldots, N$.

The proof, by induction, is straightforward and may be found in Barlow, Hunter, and Proschan (1963).

THEOREM 3.3. Let f be PF_2 with $f(x + \Delta)/f(x)$ strictly decreasing for $x \geq 0$, $\Delta > 0$, and with $f(t) > 0$ for $t > 0$. Then in the notation of Theorem 3.2 for $x_1 > x_1^*$, $\delta_n > \delta_{n-1}$ for some positive integer n; for $x_1 < x_1^*$, $\delta_n < 0$ for some positive integer n.

Proof. $F(x)/f(x)$ is strictly increasing by an obvious modification of Theorem 1 of Appendix 1. Thus an increase in x_1 from x_1^* results in an increase in δ_1. By using the lemma, and the fact that $f(x + \Delta)/f(x)$ is decreasing for fixed $\Delta > 0$, we obtain that for some n, $\delta_n > x_1$. Thus for some n, $\delta_n > \delta_{n-1}$.

Similarly, a decrease in x_1 from x_1^* results in larger and larger decreases of δ_k as k increases so long as δ_k remains positive. In fact, the decrease can be made as large as we please by taking k sufficiently large. Thus $\delta_n < 0$ for some n. ‖

We can now specify the computing procedure for obtaining the optimal inspection schedule for f satisfying the hypothesis of Theorem 3.3:

ALGORITHM 1. 1. Choose x_1 to satisfy

$$c_1 = c_2 \int_0^{x_1} (x_1 - t) \, dF(t) = c_2 \int_0^{x_1} F(t) \, dt.$$

The cost of a single check is thus balanced against the expected cost of undetected failure occurring before the first check.

2. Compute $x_2, x_3, \ldots$ recursively from (3.5).

3. If any $\delta_k > \delta_{k-1}$, reduce x_1 and repeat. If any $\delta_k < 0$, increase x_1 and repeat.

4. Continue until $x_1 < x_2 < \cdots$ are determined to the degree of accuracy required.

Example 2. Suppose the time to failure is normally distributed with mean $\alpha = 500$ hours and standard deviation $\sigma = 100$. Suppose furthermore that $c_1 = 10$ and $c_2 = 1$. Then for the first check at times $x_1 = 422.4$ hours and $x_1 = 422.5$ hours, we have the following computed times for additional checks.

k	x_k	δ_k	k	x_k	δ_k
1	422.50	64.29	1	422.40	64.20
2	486.79	.	2	486.60	.
.	.	.	.	.	.
.	.	.	.	.	.
.	.	.	.	.	.
8	.	27.48	13	.	10.50
9	727.81	27.40	14	815.98	1.00
10	755.21	30.00	15	816.98	−9.00
11	785.21		16	807.98	

Hence $x_1 = 422.5$ is too large since $\delta_{10} > \delta_9$, and $x_1 = 422.4$ is a little too small since $\delta_{15} < 0$. The proper time to schedule the first check is between 422.4 hours and 422.5 hours.

Example 3. Uniform distribution. Suppose the time to failure of a system is uniformly distributed over the interval $[0, T]$. Then the expected cost following schedule $x_1 < x_2 < \cdots$ is

$$C = \sum_{k=0}^{\infty} \int_{x_k}^{x_{k+1}} [c_1(k+1) + c_2(x_{k+1} - t)] \frac{dt}{T}.$$

By (3.5) the solution satisfies

$$x_{k+1} - x_k + \frac{c_1}{c_2} = x_k - x_{k-1}, \quad k = 2, 3, \ldots$$

Solving for x_k in terms of x_1, we obtain

$$x_k = kx_1 - \frac{k(k-1)}{2}\frac{c_1}{c_2}.$$

Hence $x_1 > \frac{k-1}{2}\frac{c_1}{c_2}$, implying that we make only a finite number of checks, say n.

Since we require $x_n = T$, we find that

$$x_k = \frac{kT}{n} + k(n-k)\frac{c_1}{2c_2}, \quad k = 0, 1, 2, \ldots, n. \tag{3.6}$$

For $\{x_k\}$ to constitute a checking schedule we must have

$$x_{k+1} - x_k > 0,$$

or $\quad \dfrac{T}{n} + \dfrac{c_1}{2c_2}(n - 2k - 1) > 0, \quad k = 1, 2, \ldots, n - 1.$

For $k = n - 1$ we obtain

$$n(n-1) < \frac{2c_2 T}{c_1}. \tag{3.7}$$

But the expected cost using $n + 1$ checks minus the expected cost using n checks is

$$-\frac{c_2}{2T}\frac{n+1}{n}\left(\frac{T}{n+1} - \frac{c_1 n}{2c_2}\right)^2 < 0.$$

Hence n must be the largest integer satisfying (3.7). For this value of n we determine $x_1 < x_2 < \cdots < x_n$ according to (3.6).

As a numerical example, suppose the time to failure is uniformly distributed over [0, 100] and $c_1 = 2$, $c_2 = 1$. Then the optimum $n = 10$ and successive checks from (3.6) should be made at times 19, 36, 51, 64, 75, 84, 91, 96, 99, 100.

Minimax Checking Schedules. Suppose now that we have no information concerning the failure distribution F. It then becomes appropriate to find the minimax checking schedule, that is, the sequence of check times $x_1 < x_2 < \cdots$ such that the maximum possible expected cost (taken over all failure distributions) is minimized by our choice of $\{x_i\}$.

Derman (1961) obtained the minimax schedule. In his model he permitted the additional generalization that failure is not necessarily detected with certainty but has probability $p > 0$ of being detected on a given inspection. His model differs further in that he assumes that in the absence of failure the unit remains in service only until some stipulated time T, rather than indefinitely as we have assumed in the model of this section.

Derman's minimax schedule when $p = 1$ corresponds to the optimal inspection schedule obtained in Example 3 when the failure distribution is the uniform on $[0, T]$.

Additional results along these lines are obtained by Roeloffs (1962). Roeloffs' model differs from Derman's in that p is taken to be 1 and in that a percentile is known of the otherwise unknown failure distribution. Under these assumptions the minimax inspection schedule is obtained.

3.2 Minimizing expected cost assuming renewal at detection of failure

In Section 3.1 we derived optimum inspection schedules under the assumption that the time span extended only until detection of failure.

OPTIMUM MAINTENANCE POLICIES 115

In many situations, however, a renewal (resulting from repair or replacement) occurs upon detection of failure. As examples, consider the following.

1. A missile is inspected from time to time to determine whether it is operationally ready. Upon detection of malfunction or failure, repair is made to return the missile to an operationally ready state.

2. A drug is stored for use during epidemics. It is tested from time to time to detect loss of potency. Upon detection of such loss, the drug is replaced. The cycle of testing, failure detection, and replacement continues indefinitely.

3. A machine produces units continuously. The quality of the output is checked at various times to determine whether the machine is functioning satisfactorily. Upon detection of malfunction, repair is made, production resumes, and inspection continues.

In such situations the optimal inspection schedule is one that minimizes expected cost per unit of time over an infinite time span. In this section we present a method suggested by David Brender (1963) for deriving such an optimal inspection schedule in certain cases.

We make assumptions (*a*) through (*e*) listed at the beginning of Section 3.1. We replace assumption (*f*) by: (*f*′) Upon detection of failure, repair (or replacement) occurs, costing c_3, on the average, and requiring time r, on the average. The unit is then as good as new and checking resumes.

If checks are made at times $x_1 < x_2 < \cdots$ following completion of repair, the expected cost per unit of time over an infinite time span $R(\mathbf{x})$ is readily computed to be

$$R(\mathbf{x}) = \frac{C(\mathbf{x})}{T(\mathbf{x})}, \tag{3.8}$$

where $\mathbf{x} = (x_1, x_2, \ldots)$,

$C(\mathbf{x})$ = expected total cost per cycle; that is, between completion of a repair and completion of the following repair, following inspection schedule $\mathbf{x}$ during each cycle,

$T(\mathbf{x})$ = expected length of a cycle following schedule $\mathbf{x}$.

Clearly,

$$C(\mathbf{x}) = \sum_{k=0}^{\infty} \int_{x_k}^{x_{k+1}} [c_1(k+1) + c_2(x_{k+1} - t)] \, dF(t) + c_3 \tag{3.9}$$

$$T(\mathbf{x}) = \mu + \sum_{k=0}^{\infty} \int_{x_k}^{x_{k+1}} (x_{k+1} - t) \, dF(t) + r. \tag{3.10}$$

Now define

$$D(\alpha, \mathbf{x}) = C(\mathbf{x}) - \alpha T(\mathbf{x}).$$

We can now specify the computing procedure for obtaining the schedule $\mathbf{x}$ which minimizes $R(\mathbf{x})$ for f satisfying the hypothesis of Theorem 3.3.

Note that $(c_1 + c_3)/(\mu + r) \leq c_2$ if the inspection is to be nontrivial, for otherwise, "perfect" inspection and repair would cost more per unit of time than performing no inspection and repair at all. By "perfect" inspection we mean detection of failure the instant it occurs, by a single check.

ALGORITHM 2. 1. For given α, find $\mathbf{x}(\alpha)$ which minimizes $D(\alpha, \mathbf{x})$, using Algorithm 1 of Section 3.1.

2. Vary α to find $\alpha = \alpha^*$ for which $D(\alpha^*, \mathbf{x}(\alpha^*)) = 0$. Such an α^* exists as shown in the proof. Then $\mathbf{x}(\alpha^*)$ minimizes $R(\mathbf{x})$.

Proof. Note that $D(c_2, \mathbf{x}(c_2)) = c_1 + c_3 - c_2(\mu + r) \leq 0$. At the same time $D(0, \mathbf{x}(0)) > 0$. Since $D(\alpha, \mathbf{x}(\alpha))$ can be shown continuous in α for $\alpha \leq c_2$, it follows that $\alpha = \alpha^*$ exists for which $D(\alpha^*, \mathbf{x}(\alpha^*)) = 0$. By definition of $\mathbf{x}(\alpha^*)$, for any $\mathbf{x}$

$$D(\alpha^*, \mathbf{x}(\alpha^*)) \leq D(\alpha^*, \mathbf{x}).$$

Since $D(\alpha^*, \mathbf{x}(\alpha^*)) = 0$ and $D(\alpha^*, \mathbf{x}) = C(\mathbf{x}) - \alpha^* T(\mathbf{x})$, it follows that

$$R(\mathbf{x}) = \frac{C(\mathbf{x})}{T(\mathbf{x})} \geq \alpha^*.$$

But $D(\alpha^*, \mathbf{x}(\alpha^*)) = 0$ implies

$$\alpha^* = \frac{C(\mathbf{x}(\alpha^*))}{T(\mathbf{x}(\alpha^*))}.$$

Hence for all $\mathbf{x}$

$$\frac{C(\mathbf{x})}{T(\mathbf{x})} \geq \frac{C(\mathbf{x}(\alpha^*))}{T(\mathbf{x}(\alpha^*))},$$

so that $\mathbf{x}(\alpha^*)$ minimizes $R(\mathbf{x})$. ∥

If f does not satisfy the hypothesis of Theorem 3.3, the solution requires a slightly different approach, which we briefly sketch. We write

$$D(\alpha, \mathbf{x}) = C(\mathbf{x}) - \alpha T(\mathbf{x})$$
$$= \sum_{k=0}^{\infty} \int_{x_k}^{x_{k+1}} [c_1(k+1) + (c_2 - \alpha)(x_{k+1} - t)] \, dF(t) + c_3 - \alpha(\mu + r).$$

A necessary condition for a minimum for $D(\alpha, \mathbf{x})$ as $\mathbf{x}$ is varied is obtained by setting each $\partial D(\alpha, \mathbf{x})/\partial x_k = 0$, yielding

$$x_{k+1} = x_k + \frac{F(x_k) - F(x_{k-1})}{f(x_k)} - \frac{c_1}{c_2 - \alpha}, \qquad k = 2, 3, \ldots. \qquad (3.11)$$

Once x_1 is selected, $x_2, x_3, \ldots$ may be obtained recursively from (3.11), and $D(\alpha, \mathbf{x})$ calculated. By varying x_1 we find the sequence $\mathbf{x}(\alpha)$ which achieves the minimum value for $D(\alpha, \mathbf{x})$. As before we then repeat this procedure for different α until we find α^* for which $D(\alpha^*, \mathbf{x}(\alpha^*)) = 0$. Then $\mathbf{x}(\alpha^*)$ minimizes $R(\mathbf{x})$. See Brender (1963) for details when $r = 0$.

3.3 Opportunistic replacement of a single part in the presence of several monitored parts

In all the maintenance models studied thus far, we have assumed only a one-unit system. In this section we consider more complex systems, so that a new aspect of the maintenance problem is introduced; namely, the maintenance action to be taken on one part is made to depend on the state of the rest of the system. This section is based on the work of Jorgenson and Radner (1962) and Radner and Jorgenson (1963). See Chapter 5, Section 5, for another approach to such maintenance problems.

In Jorgenson and Radner (1962), the following model is considered. A system consists of parts 0 and 1 in series having independent failure distributions F_0 and F_1 respectively, where $F_1(t) = 1 - e^{-\lambda t}$. Part 1 is continuously monitored so that its state (operating or failed) is known at every moment of time. Part 0 cannot be inspected except possibly at the time of replacement, so that its state is in general not known. Replacement of part i takes K_i units of time, $i = 0, 1$; replacement of parts 0 and 1 together takes K_{01} units of time, where $K_0, K_1 \leq K_{01} \leq K_0 + K_1$. The objective function to be maximized is the expected total discounted time that the system is in a good state; discounting good time takes into account the fact that a unit of good time at the present is worth more than a corresponding unit in the future. Specifically, if $(t, t + dt)$ is an interval of good time, its contribution toward total good time is $e^{-\alpha t} dt + o(dt)$, where $0 \leq \alpha < \infty$ is called the discount factor. If the time span is infinite, the objective function is the long-run average proportion of the time the system is good.

Using a dynamic programming formulation, Jorgenson and Radner show that the optimum replacement policy is what they call an (n, N) type of policy: (1) If part 1 fails when the age of part 0 is between 0 and n, replace part 1 alone. (2) If part 1 fails when the age of part 0 is between n and N, replace parts 0 and 1 together. (3) If part 1 has not yet failed when the age of part 0 is N, replace part 0 alone.

As Jorgenson and Radner point out, the (n, N) policy plays the same role in the field of inspection as the (s, S) policy plays in the field of inventory. Thus the difference $K_0 + K_1 - K_{01}$ plays the same role as setup cost or order cost in inventory theory. As $K_0 + K_1 - K_{01} \to 0$, $N - n \to 0$; this parallels the relationship in inventory theory in which as the order cost approaches zero, $S - s \to 0$ (Karlin, 1958, p. 233). The (n, N) policy has two advantages: (1) it is simple to administer, and (2) the optimal values of the parameters n and N are easy to compute.

A generalization of the model is made by Radner and Jorgenson (1962). The system now has $M + 1$ parts in series, with independent failure distributions F_i $(i = 0, 1, \ldots, M)$ where $F_i(t) = 1 - e^{-\lambda_i t}$, $(i = 1, 2, \ldots, M)$, whereas the only assumption on $F_0(t)$ is that it has positive density for all positive t and that $F_0(0) = 0$. Parts $1, 2, \ldots, M$ are continuously monitored assuring immediate determination of any failure. Part 0 cannot be inspected except possibly at the time of replacement. Replacement of part i alone takes time K_i and costs C_i $(i = 0, 1, \ldots, M)$. Replacement of parts 0 and i together takes time K_{0i} and costs C_{0i} $(i = 1, 2, \ldots, M)$ where K_0, $K_i \leq K_{0i} \leq K_0 + K_i$ and C_0, $C_i \leq C_{0i} \leq C_i + e^{-\alpha K_i} C_0$ $(i = 1, 2, \ldots, M)$, and α is the discount factor. The unit of time is chosen so that one unit of good time is worth one unit of money.

Again using a dynamic programming formulation, Radner and Jorgenson show that the optimum replacement policy is of the type that they call an (n_i, N) policy: (1) If part i fails when the age of part 0 is between 0 and n_i, replace part i alone $(i = 1, 2, \ldots, M)$. (2) If part i fails when the age of part 0 is between n_i and N, replace parts 0 and i together. (3) If parts $1, 2, \ldots, M$ have all survived until the age of part 0 is N, replace part 0 alone. The policy is optimum with respect to each of the following objective functions: expected total discounted value of good time minus cost, expected total discounted time in which the system is in a good state, ratio of expected total discounted good time to expected total discounted cost. For an infinite time span, the expected total discounted sums are replaced by long-run averages per unit of time.

CHAPTER 5

Stochastic Models for Complex Systems

1. INTRODUCTION

In previous chapters we have considered only systems which can assume at most two or three possible states. In this chapter we consider more complex systems which are subject to certain inspection-repair-replacement policies. At any instant in time the system (a given equipment configuration) can be in one of a number of possible states. A state can often be defined simply by listing the equipments which are working satisfactorily. In general, the number of distinguished states depends on the number and function of the system equipments. In the models that we shall consider, the number of possible states is considered finite.

The deterioration law of the system will usually be assumed to be Markovian; that is, the future course of the system depends only on its state at the present time and not on its past history. This model has been suggested by Klein (1962), among others. There are at least two good reasons for suggesting a Markov model to describe deterioration. First of all, if each component has approximately an exponential failure law, the complete system can be described approximately by a Markov process. Secondly, the first order approximating description of many physical systems is that in which knowledge of the history of such systems contains no predictive value. A Markov process is the stochastic equivalent of this type of process.

To illustrate how the state space and flow diagram for a complex system may be constructed, we consider a model based on the operation of a large-scale warning system. This example will subsequently be used to illustrate the techniques discussed.

Example 1.1. A key component of a certain radar warning system consists of two identical computers connected in parallel; that is, both are operating although only one is in actual service. Emergency repair is performed at the time of a computer failure. Preventive maintenance for

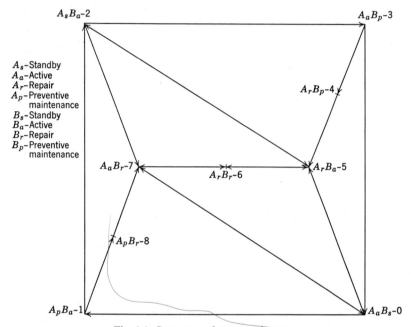

Fig. 1.1. State space for two-unit system.

a specified computer is scheduled after t_0 hours if one computer is active and the other is on an operating standby basis. If the first computer fails (is undergoing preventive maintenance) and the second fails before the first is repaired (preventive maintenance is completed), the consequences could be catastrophic because of the nature of the system. For convenience, we denote the possible states for the system as follows. Label one computer A and the other B. Then

$A_s(B_s)$ denotes that computer $A(B)$ is in an operating *standby* state;
$A_a(B_a)$ denotes that computer $A(B)$ is in an *active* state;
$A_r(B_r)$ denotes that computer $A(B)$ is undergoing *emergency repair*;
$A_p(B_p)$ denotes that computer $A(B)$ is undergoing *preventive maintenance*.

Figure 1.1 is a diagram of the state space of the system. There are altogether nine possible states for the system, labeled 0 through 8 respectively. For example, if the system is in state 0, computer A is being used actively while computer B is on an operating standby basis. If no failures occur in a time interval of length t_0 measured from the moment we enter state 0, preventive maintenance is performed on computer A and we enter state 1. If no failures occur, the state of the system passes around the perimeter of the square, so to speak. If the active computer fails while the second computer is undergoing preventive maintenance, the system is of course down. There are exactly three unfavorable states so far as system operation is concerned, namely states 4, 6, and 8. Under certain reasonable assumptions on the time to failure, the time to perform repair, etc., the operation of the system can be described by a so-called semi-Markov process (see Section 2.2). The user of this system may be interested in the mean system down time during a specified time interval, the probability that the system is down more than x minutes at any one time, or perhaps an appropriate maintenance schedule.

In order to obtain such information, we shall develop some of the key ideas of Markov chains and semi-Markov processes in Section 2. In Section 3 selected repairman-type problems of widespread interest are discussed. In Section 4 we treat marginal checking and obtain qualitative information about optimal replacement rules for systems which can be described by IFR Markov chains. In Section 5 we continue the discussion of optimal control rules for Markovian systems, reducing optimization to a problem in linear programming.

2. MARKOV CHAINS AND SEMI-MARKOV PROCESSES

We confine attention to Markov chains having a finite state space and either a discrete or continuous time parameter. We shall also discuss a more general process called a "semi-Markov process." This process has an embedded discrete parameter Markov chain and represents, figuratively speaking, a marriage of renewal theory (see Section 2 of Chapter 3) and Markov chain theory. These processes will be used to describe certain classes of systems subject to inspection, replacement, and repair.

A discrete parameter stochastic process $\{X(t);\ t = 0, 1, \ldots\}$ or a continuous parameter process $\{X(t);\ t \geq 0\}$ is said to be a *Markov process* if, for any set of n time points $t_1 < t_2 < \cdots < t_n$ in the index set of the process and any real numbers $x_1, x_2, \ldots, x_n$,

$$P[X(t_n) \leq x_n \mid X(t_1) = x_1, \ldots, X(t_n) = x_{n-1}]$$
$$= P[X(t_n) \leq x_n \mid X(t_{n-1}) = x_{n-1}]. \quad (2.1)$$

Intuitively, (2.1) means that given the present state of the process, the future states of the process are independent of its past.

Discrete time Markov chains

A discrete time Markov chain is described by a sequence of discrete-valued random variables $\{X(t_n)\}_{n=1}^{\infty}$. Without loss of generality we can identify the times of state transitions by $n = 0, 1, 2, \ldots$ and the state at time n by $X(n)$. It will be convenient to label the states of the process by the nonnegative integers $i = 0, 1, 2, \ldots, m$. Thus a transition from i to j means a change from the state labeled i to the state labeled j.

A Markov chain is determined when the one-step transition probabilities of the state variables are specified; that is, a conditional probability (called the transition probability) of a transition at time n for each pair i, $j = 0, 1, 2, \ldots, m$ from state i to state j must be given. This probability is denoted by

$$p_{ij}^{n,n+1} = P[X(n+1) = j \mid X(n) = i].$$

When the transition probability functions depend only on the time difference, that is,

$$p_{ij}^{n,n+1} = p_{ij}^{0,1} = p_{ij}$$

we say that the Markov process is *stationary* in time. Henceforth we confine ourselves to time stationary Markov chains. The initial state of the process must also be prescribed. If we allow the initial state to be random, we suppose that the distribution function of the initial state is given. It is customary to arrange the numbers p_{ij} as a matrix and refer to $P = (p_{ij})$ as the Markov transition probability matrix of the process. Clearly, the values p_{ij} satisfy

$$p_{ij} \geq 0, \qquad i, j = 0, 1, \ldots, m$$

$$\sum_{j=0}^{m} p_{ij} = 1.$$

All the quantities which we will be interested in, such as the expected number of steps to go from state i to state j for the first time, the expected number of occurrences of state j in n steps, etc., can be computed from matrices consisting of functions of P. A complete treatment of such processes, using only elementary mathematics, can be found in Kemeny and Snell (1960).

Example 2.1. In Example 1.1 we described a stochastic process generated by the succession of repairs, preventive overhauls, etc., of a certain system. There were $m + 1 = 9$ states in all, and t_0 denoted the scheduled time between preventive overhauls. Suppose that the repair distribution G is exponential, that is,

$$G(t) = \begin{cases} 0, & t < 0 \\ 1 - e^{-\theta t}, & t \geq 0. \end{cases}$$

An electronic computer may be susceptible to failures which can be detected while the machine is running and also to failures which can be detected only by shutting the computer down and examining it very closely. We suppose that the failure distribution for the first type of failure is exponential, that is,

$$F(t) = \begin{cases} 0, & t < 0 \\ 1 - e^{-\lambda t}, & t \geq 0. \end{cases}$$

Scheduled preventive maintenance is desirable to catch failures which may not be detectable while the machine is running.

The process has an embedded Markov chain with probability transition matrix given in Figure 2.1. For example, $p_{00} = 0$ since the process must leave the first state through either a failure or a preventive overhaul. Moreover, $p_{01} = e^{-2\lambda t_0}$ is the probability that the process goes from state 0 directly to state 1, that is, that there is no failure in $[0, t_0]$. From state 4 (8) the process must go directly to state 5 (7) assuming that a computer undergoing preventive maintenance is never more than d minutes away from operation where d is short.

Classification of states of a Markov chain

Two states i and j are said to *communicate*, that is, $i \sim j$, if and only if there exist integers $n_{ij} > 0$ and $n_{ji} > 0$ such that $p_{ij}^{n_{ij}} > 0$ and $p_{ji}^{n_{ji}} > 0$, where p_{ij}^n is the probability that starting in state i we will be in state j after n steps. An *ergodic* set of states is one in which all states communicate and which cannot be left once it is entered. An *ergodic state* is an element of an ergodic set. If we examine the original process which generated the transition matrix in Figure 2.1, we can easily convince ourselves that the set of states is ergodic. An *absorbing* state is a one-element ergodic set; that is, i is an absorbing state if and only if $p_{ii} = 1$. A state is called *transient* if it is not ergodic.

We define the *period* of a state i, written $d(i)$, to be the greatest common divisor of all integers $n \geq 1$ for which $p_{ii}^n > 0$. If $d(i) = 1$, the state i is said to be *nonperiodic*. It is not too hard to show that if $i \sim j$, then $d(i) = d(j)$. Hence if a Markov chain is ergodic (that is, all states communicate), then to show that it is nonperiodic it is enough to show that, say, $d(0) = 1$.

$\frac{1}{\lambda}$ —Mean time to failure

$\frac{1}{\theta}$ —Mean time to perform emergency repair

γ —Time to perform preventive maintenance

t_0 —Scheduled preventive maintenance period

d —Switchover time when one unit is in preventive maintenance

	0	1	2	3	4	5	6	7	8
0	0	$e^{-2\lambda t_0}$	0	0	0	$\frac{1-e^{-2\lambda t_0}}{2}$	0	$\frac{1-e^{-2\lambda t_0}}{2}$	0
1	0	0	$e^{-\lambda\gamma}$	0	0	0	0	0	$1-e^{-\lambda\gamma}$
2	0	0	0	$e^{-2\lambda t_0}$	0	$\frac{1-e^{-2\lambda t_0}}{2}$	0	$\frac{1-e^{-2\lambda t_0}}{2}$	0
3	$e^{-\lambda\gamma}$	0	0	0	$1-e^{-\lambda\gamma}$	0	0	0	0
4	0	0	0	0	0	1	0	0	0
5	0	0	$\frac{\theta}{\lambda+\theta}$	0	0	0	$\frac{\lambda}{\lambda+\theta}$	0	0
6	0	0	0	0	0	$\frac{1}{2}$	0	$\frac{1}{2}$	0
7	$\frac{\theta}{\lambda+\theta}$	0	0	0	0	0	$\frac{\lambda}{\lambda+\theta}$	0	0
8	0	0	0	0	0	0	0	1	0

Fig. 2.1. Transition matrix.

Absorbing Markov chains

An absorbing chain is one in which all the nontransient states are absorbing. The transition probability matrix of an absorbing Markov chain with s transient states and $r - s$ absorbing states can be expressed in the following canonical form

$$P = \left(\begin{array}{c|c} \overset{r-s}{I} & \overset{s}{0} \\ \hline R & Q \end{array} \right) \begin{array}{c} r-s \\ s \end{array}$$

where I is the $r - s$ by $r - s$ identity matrix and 0 is the $r - s$ by s zero matrix. Kemeny and Snell (1960) define the fundamental matrix for an absorbing Markov chain to be

$$N = (I - Q)^{-1}.$$

If we define N_{ij} to be the total number of times the process is in state j before absorption if it starts in state i and

$$E[N_{ij}] = n_{ij}$$

to be the expected number of visits to state j before absorption starting in state i, then

$$N = (n_{ij}) \qquad (2.2)$$

(Kemeny and Snell, 1960, p. 46). Furthermore

$$E[N_{ij}^2] = N(2N_{dg} - I) \qquad (2.3)$$

where N_{dg} is formed by replacing the off-diagonal elements of N by zero. It can be shown that (2.3) and hence (2.2) are always finite. Kemeny and Snell note that it is often the case that Var $[N_{ij}]$ is quite large compared to $\{E[N_{ij}]\}^2$; hence the means are fairly unreliable estimates for Markov chains.

We can also use (2.2) quite effectively for Markov chains in which all states communicate. Suppose we are interested in the expected number of visits to state j starting in state i before reaching state k. We can convert state k to an absorbing state by making $p_{kk} = 1$ and $p_{kj} = 0$ for $j \neq k$. This converts our original Markov chain to an absorbing Markov chain with absorbing state k, and n_{ij} for this converted chain is exactly the required quantity.

In Example 2.1 we might require the expected number of visits to state j starting in state i before the system entered the state of complete failure—state 6. Suppose the mean time to failure $1/\lambda$ equals 35 hours and the

126   MATHEMATICAL THEORY OF RELIABILITY

$1/\lambda = 35$ hours
$1/\theta =$ hour
$\gamma = 1$ hour
$t_0 = 24$ hours

	0	1	2	3	4	5	6	7	8
0	0	.268667	0	0	0	.365667	0	.365667	0
1	0	0	.971833	0	0	0	0	0	.028167
2	0	0	0	.268667	0	.365667	0	.365667	0
3	.971833	0	0	0	.028167	0	0	0	0
$P=4$	0	0	0	0	0	1	0	0	0
5	0	0	.995261	0	0	0	.004739	0	0
6	0	0	0	0	0	$\tfrac{1}{2}$	0	$\tfrac{1}{2}$	0
7	.995261	0	0	0	0	0	.004739	0	0
8	0	0	0	0	0	0	0	1	0

mean repair time $1/\theta$ equals 10 minutes. Suppose, furthermore, preventive maintenance is scheduled every 24 hours on the computer operating at the time of scheduling. Assume also that the time to actually perform preventive maintenance is a constant, say γ. The transition probability matrix becomes that shown in Figure 2.2. Striking out row 6 and column 6 of $I - P$ and inverting, we obtain n_{ij}, the expected number of visits to state j before complete failure. In Figure 2.3 we see that the system is in state 5 about 105 times compared to only 38 times for state 1 before a complete failure occurs. This could have been anticipated, however, since the mean time to first failure with two computers is $17\frac{1}{2}$ hours, and we have scheduled preventive maintenance only every 24 hours. The expected *actual time* during which the system is in states 0, 1, 2, 3 before entering state 6 is much greater than the expected actual time that the system is in states 4, 5, 7, and 8, however. This can be verified by using Theorem 2.5.

It is also easy to compute the probability h_{ij} that a Markov chain process will ever go to transient state j, starting in transient state i. The probability is given by the element in the ith row, jth column of

$$H = (h_{ij}) = (N - I)N_{dg}^{-1}$$

(see Kemeny and Snell, 1960, p. 61).

Ergodic Markov chains

An ergodic Markov chain is one in which all states communicate and hence all have the same period, say d. As in our discussion of absorbing Markov chains, the matrix $I - P$ plays a leading role.

Let

$$u_j(n) = \begin{cases} 1, & \text{if } X(n) = j \\ 0, & \text{otherwise.} \end{cases}$$

Then the expected number of visits to state j in n steps if the process starts in state i is

$$E_i\left[\sum_{k=1}^{n} u_j(k)\right] = \sum_{k=1}^{n} p_{ij}^k.$$

This quantity can be used to characterize ergodic and transient states.

THEOREM 2.1. A state i of a Markov chain is ergodic (transient) if and only if $\sum_{k=1}^{\infty} p_{ii}^k = +\infty \ (<\infty)$.

For a proof see Parzen (1962, p. 216).

	0	1	2	3	4	5	7	8
0	143.192	38.4708	142.394	38.2565	1.07758	105.507	105.513	1.08363
1	142.395	39.2567	143.148	38.4590	1.08349	105.497	105.519	1.10575
2	142.392	38.2558	143.189	38.4701	1.08359	105.511	105.505	1.07755
$N = 3$	143.148	38.4589	142.395	39.2567	1.10575	105.519	105.497	1.08327
4	141.717	38.0846	142.511	38.2879	2.07846	106.011	105.005	1.07245
5	141.717	38.0746	142.511	38.2879	1.07846	106.011	105.005	1.07245
7	142.511	38.2878	141.717	38.0745	1.07245	105.005	106.011	1.07845
8	142.511	38.2878	141.717	38.0745	1.07245	105.005	106.011	2.07845

Fig. 2.3. N matrix with state 6 absorbing.

When P is ergodic, we can produce a solution vector $\pi = (\pi_0, \pi_1, \ldots, \pi_m)$ such that $\pi P = \pi$ in terms of subdeterminants of $I - P$. Let D_i denote the determinant formed by striking out the ith row and the ith column of $I - P$.

THEOREM 2.2. A Markov chain with probability transition matrix P is ergodic if and only if $D_i > 0$ for $i = 0, 1, \ldots, m$. If $\pi = (D_0, D_1, \ldots, D_m)$, then $\pi P = \pi$.

This theorem is due to Mihoc and can be found in Fréchet (1950, vol. 2, pp. 114–116).

It will be convenient to let $\xi' = (1, 1, \ldots, 1)$ denote a row vector of ones. Note that $\xi\pi$ is a matrix when π is a row vector.

THEOREM 2.3. If P is an ergodic probability transition matrix with period d and $\pi = (\pi_0, \pi_1, \ldots, \pi_m)$ where $\pi_i = D_i \Big/ \sum_{j=0}^{m} D_j$, then
(a) $\lim_{n \to \infty} P^{nd} = \xi\pi = \Pi$,
(b) $P\Pi = \Pi P = \Pi$,
(c) π is the unique probability vector satisfying $\pi P = \pi$.

Proof. (a) is the standard ergodic theorem for Markov chains (Kemeny and Snell, 1960, p. 71). The uniqueness of the fixed probability vector in (c) follows immediately from (a), (b), and Theorem 2.2. ∥

If the Markov chain is nonperiodic, that is, $d = 1$, then Theorem 2.3 says

$$\lim_{n \to \infty} p_{ij}^n = \pi_j = \frac{D_j}{\sum_{k=0}^{m} D_k}.$$

Note that π_j is the probability of being in state j after an infinite number of transitions have occurred. It can be shown (Kemeny and Snell, 1960) that there exist constants b and r with $0 < r < 1$ such that $p_{ij}^n = \pi_j + e_{ij}^n$ with $|e_{ij}^n| < br^n$; that is, p_{ij}^n approaches its limit geometrically fast. We shall often be interested in the stationary probability vector π.

Example 2.2. In Figure 2.1 we tabulated the transition matrix for a system subject to preventive maintenance and repair at failure. The vector of stationary probabilities can easily be computed for this process. Noting the natural symmetry of the state space in Figure 1.1, we know that

$$\pi_0 = \pi_2$$
$$\pi_1 = \pi_3$$
$$\pi_7 = \pi_5$$
$$\pi_8 = \pi_4.$$

Because of the profusion of zero entries in P, these stationary probabilities can be determined without recourse to determinants by using the equation $\pi P = \pi$. They are

$$\pi_1 = e^{-2\lambda t_0}\pi_0$$

$$\pi_5 = \frac{(\lambda + \theta)}{\theta}(1 - e^{-\lambda(2t_0+\gamma)})\pi_0$$

$$\pi_4 = e^{-2\lambda t_0}(1 - e^{-\lambda\gamma})\pi_0$$

$$\pi_6 = \frac{2\lambda}{\theta}(1 - e^{-\lambda(2t_0+\gamma)})\pi_0.$$

Using the fact that $\sum_{k=0}^{8} \pi_k = 1$ we can determine π_0 and hence all stationary probabilities.

A quantity of some interest is the number of steps required to go to state j for the first time from state i. This is the first-passage time and is denoted by the random variable Y_{ij}. Let $M = (m_{ij})$ where $m_{ij} = E[Y_{ij}]$ is the mean first-passage time. It is readily verified that the m_{ij} satisfy a renewal-type equation, namely,

$$m_{ij} = \sum_{k \neq j} p_{ik}(m_{kj} + 1) + p_{ij}, \tag{2.4}$$

or in matrix notation

$$M = P(M - M_{dg}) + E \tag{2.5}$$

where M_{dg} is formed by replacing off-diagonal elements of M by zeros and E is the unit matrix all of whose entries are one. Multiplying (2.5) by π we obtain

$$\pi M_{dg} = \pi E$$

or

$$m_{ii} = 1/\pi_i.$$

To solve for the matrix M, Kemeny and Snell introduce the fundamental matrix for ergodic Markov chains

$$Z = [I - (P - \Pi)]^{-1}.$$

It can be shown that this matrix always exists for ergodic Markov chains and that M is given by

$$M = (I - Z + EZ_{dg})D, \tag{2.6}$$

where $D = (d_{ij})$ and $d_{ij} = 0$, $i \neq j$, $d_{ii} = 1/\pi_i$.

The variance of the first-passage time can also be explicitly computed. Let

$$W = \{E[Y_{ij}^2]\}$$

then

$$W = M(2Z_{dg} D - I) + 2[ZM - E(ZM)_{dg}].$$

It can be shown that W and hence M is always finite. Furthermore,

$E[Y_{ij}^2]$ is often large compared to $\{E[Y_{ij}]\}^2$, indicating that the mean first-passage times are probably unreliable estimates of first-passage times.

Another quantity of interest is the renewal quantity. Let $N_{ij}(n)$ denote the number of visits to state j in n steps if the process starts in state i. Let

$$\gamma_{ij}(n) = E[N_{ij}(n)].$$

The following theorem can be found in Kemeny and Snell (1960).

THEOREM 2.4. If P is ergodic, then

(a) $$\lim_{n \to \infty} \frac{\gamma_{ij}(n)}{n} = \pi_j$$

(b) $$(\gamma_{ij}(n) - n\pi_j) \to \pi(Z - \Pi)$$

as $n \to \infty$.

Another quantity of some interest is the mean number of visits to state j between occurrences of state i. It can be shown that this quantity is given by π_j/π_i.

Semi-Markov processes

The Markov chain process can be generalized in a natural way to a semi-Markov process as follows. As before, let $P = (p_{ij})$ denote the transition matrix of a time homogeneous Markov chain with $m + 1$ states (that is, $i, j = 0, 1, \ldots, m$).

We shall heuristically define a stochastic process $\{Z(t), t \geq 0\}$, where $Z(t) = i$ denotes that the process is in state i at time t. Given that the process has just entered a state, say i, the selection of the next state is made according to the matrix $P = (p_{ij})$. The distribution function for the "wait" of the process in state i given that the next transition will be to state j is denoted by $F_{ij}(t)$. Let $\mathscr{F}(t) = (F_{ij}(t))$. The process is Markovian only at certain "Markov points" in time at which the state transitions take place. If we specify the vector of initial probabilities $(a_0, a_1, \ldots, a_m)$, the resulting process is called a semi-Markov process. If we let $N_j(t)$ denote the number of times in $[0, t]$ that the process is in state j, the family of vectors

$$[N_0(t), N_1(t), \ldots, N_m(t)], \quad t \geq 0$$

is called a *Markov renewal process* (Pyke, 1961a). The reader is referred to Pyke (1961a and 1961b) for a careful definition and discussion of these processes.

A time homogeneous Markov chain is a semi-Markov process where

$$F_{ij}(x) = \begin{array}{ll} 0, & x < 1 \\ 1, & x \geq 1. \end{array}$$

A stable, continuous time parameter Markov process is a semi-Markov process in which all waiting time distributions are exponential; that is,

$$F_{ij}(t) = 1 - e^{-\lambda_i t}, \quad t \geq 0$$

for constants $\lambda_i > 0$ for every i. If there is only one state, a Markov renewal process becomes a renewal process (see Section 2 of Chapter 3).

In connection with reliability applications we are interested in such questions as:

(i) What is the mean recurrence time to state i?
(ii) What is the expected number of occurrences of state i in time t?
(iii) What is the limiting probability that the process will be in state i?

Much of the discussion that follows can be found in Barlow (1962b).

Moments of first-passage distributions for semi-Markov processes

Let $G_{ij}(t)$ denote the distribution function for first passage measured from the moment the process enters state i until it first enters state j. A matrix equation for the Laplace transforms of first-passage distributions can be found in Pyke (1961b). Since we shall need the first and second moments, we now compute them directly.

Let $F_i(t) = \sum_{j=0}^{m} p_{ij} F_{ij}(t)$ denote the distribution function of the unconditional wait in state i, μ_i denote the unconditional mean wait in state i, and

$$M = \begin{pmatrix} \mu_1 & \mu_1 & \cdots & \mu_1 \\ \mu_2 & \mu_2 & \cdots & \mu_2 \\ \vdots & & & \\ \mu_m & \mu_m & \cdots & \mu_m \end{pmatrix}.$$

Let l_{ij} denote the mean first-passage time to go from state i to state j. Then

$$l_{ij} = \sum_{k \neq j} p_{ik}(\mu_{ik} + l_{kj}) + p_{ij}\mu_{ij},$$

where μ_{ij} is the mean of the distribution function $F_{ij}(t)$. Since

$$\mu_i = \sum_{k=0}^{m} p_{ik}\mu_{ik}$$

we obtain
$$l_{ij} = \sum_{k \neq j} p_{ik} l_{kj} + \mu_i \tag{2.7}$$

or
$$L = (l_{ij}) = P[L - L_{dg}] + M$$

where L_{dg} is obtained from L by replacing off-diagonal elements by zero.

STOCHASTIC MODELS FOR COMPLEX SYSTEMS 133

If all states communicate, there exists a vector $\pi = (\pi_0, \pi_1, \ldots, \pi_m)$ of stationary probabilities for the embedded Markov chain. Write $L = L_{dg} + L_0$ and $L_{dg} = (P - I)L_0 + M$. Applying π to both sides, we obtain $\pi L_{dg} = \pi M$, or

$$l_{ii} = \frac{1}{\pi_i} \sum_{k=0}^{m} \pi_k \mu_k. \tag{2.8}$$

This is a very important quantity and will have further uses.

Example 2.3. From the description of system operation in Example 2.1 it can be verified that

$$F_0(x) = F_2(x) = \begin{cases} 0, & x < 0 \\ 1 - e^{-2\lambda x}, & 0 \leq x < t_0 \\ 1, & x \geq t_0 \end{cases}$$

$$F_1(x) = F_3(x) = \begin{cases} 0, & x < 0 \\ 1 - e^{-\lambda x}, & 0 \leq x < \gamma \\ 1, & x \geq \gamma \end{cases}$$

$$F_4(x) = F_8(x) = \begin{cases} 0, & x < d \\ 1, & x \geq d \end{cases}$$

$$F_5(x) = F_7(x) = \begin{cases} 0, & x < 0 \\ 1 - e^{-(\lambda+\theta)x}, & x \geq 0 \end{cases}$$

$$F_6(x) = \begin{cases} 0, & x < 0 \\ 1 - e^{-2\theta x}, & x \geq 0 \end{cases}$$

and

$$\mu_0 = \mu_2 = \frac{1 - e^{-2\lambda t_0}}{2\lambda}$$

$$\mu_1 = \mu_3 = \frac{1 - e^{-\lambda \gamma}}{\lambda}$$

$$\mu_4 = \mu_8 = d$$

$$\mu_5 = \mu_7 = \frac{1}{\lambda + \theta}$$

$$\mu_6 = \frac{1}{2\theta}.$$

Using the results of Example 2.2 and (2.8) we can compute the mean recurrence time to state 0, namely,

$$l_{00} = \frac{1}{\pi_0} \sum_{k=0}^{8} \pi_k \mu_k.$$

We can also obtain a recursion relation for the second moments $l_{ij}^{(2)}$ of the first-passage distributions. Let Y_{ij} be a random variable corresponding to the first-passage distribution G_{ij}, and let X_{ij} be a random variable with distribution F_{ij}. Then

$$l_{ij}^{(2)} = \sum_{k \neq j} p_{ij} E[X_{ik} + Y_{kj}]^2 + p_{ij} E[X_{ij}]^2$$

$$= \mu_i^{(2)} + \sum_{k \neq j} p_{ik}[l_{kj}^{(2)} + 2\mu_{ik}l_{kj}].$$

Thus $$W = (l_{ij}^{(2)}) = P[W - W_{dg}] + V + 2UL_0 \qquad (2.9)$$

where

$$U = \begin{pmatrix} p_{00}\mu_{00} & \cdots & p_{0m}\mu_{0m} \\ p_{10}\mu_{10} & \cdots & p_{1m}\mu_{1m} \\ \cdots\cdots\cdots\cdots\cdots\cdots \\ \cdots\cdots\cdots\cdots\cdots\cdots \\ p_{m0}\mu_{m0} & \cdots & p_{mm}\mu_{mm} \end{pmatrix}$$

and

$$V = \begin{pmatrix} \mu_0^{(2)} & \cdots & \mu_0^{(2)} \\ \mu_1^{(2)} & \cdots & \mu_1^{(2)} \\ \cdots\cdots\cdots\cdots \\ \mu_m^{(2)} & \cdots & \mu_m^{(2)} \end{pmatrix}$$

If all states communicate, we can apply $\pi = (\pi_0, \pi_1, \ldots, \pi_m)$ to both sides of (2.9) to obtain

$$l_{ii}^{(2)} = \frac{1}{\pi_i} \left(\sum_k \pi_k \mu_k^{(2)} + 2 \sum_{k \neq i} \sum_s \pi_s p_{sk} \mu_{sk} l_{ki} \right). \qquad (2.10)$$

For a Markov chain, (2.10) reduces to

$$l_{ii}^{(2)} = \frac{1}{\pi_i} \left(1 + 2 \sum_{k \neq i} \pi_k l_{ki}' \right).$$

We will obtain a simplified version of (2.10) for continuous time parameter Markov processes.

A semi-Markov process $[P, \mathscr{F}(t)]$ is *absorbing* (*ergodic*) if the underlying Markov chain defined by P is absorbing (ergodic) and $\mu_i < \infty$ ($i = 0, 1, \ldots, m$). If the Markov chain is composed solely of transient states and k absorbing states, then by appropriate relabeling we can partition our transition matrix P into absorbing and nonabsorbing classes as follows:

$$P = \left(\begin{array}{c|c} I & 0 \\ \hline R & Q \end{array} \right) \qquad (2.11)$$

where Q is an $(n - k) \times (n - k)$ submatrix of P. From (2.2) we know that the i, jth element of

$$(I - Q)^{-1}$$

is the expected number of visits to state j starting in state i before absorption.

THEOREM 2.5. Let $[P, \mathscr{F}(t)]$ be an absorbing semi-Markov process with k absorbing states $(0, 1, \ldots, k - 1)$ where P is in the normalized form (2.11). The mean time to absorption starting in state i ($i > k$) is

$$\sum_{j=k}^{m} m_{ij}\mu_j$$

where $(m_{ij}) = (I - Q)^{-1}$.

Proof. Let $X_n^{(j)}$ denote the unconditional wait in state j on the nth visit to state j. Then $\{X_n^{(j)}\}_{n=1}^{\infty}$ are mutually independent and, furthermore, are independent of N_{ij}, the number of visits to j starting in i before absorption. Hence

$$E[X_1^{(j)} + X_2^{(j)} + \cdots + X_{N_{ij}}^{(j)}] = E[X_n^{(j)}]E[N_{ij}] = \mu_j m_{ij}$$

by Theorem 2.4, Chapter 3. ∥

Note that if all states communicate, we can modify P so that one state becomes an absorbing state. In this way the mean first-passage times can be calculated for any finite semi-Markov process.

Limit theorems

We shall be concerned with the asymptotic properties of several quantities of interest related to a finite semi-Markov process. We assume nonlattice distributions. Let $\{Z(t), t \geq 0\}$ denote a semi-Markov process and $P_{ij}(t) = P[Z(t) = j \mid Z(0) = i]$. In particular, we need the limiting value of $P_{ij}(t)$ and $\gamma_{ij}(t)$, where $\gamma_{ij}(t)$ denotes the mean number of occurrences during $[0, t]$ of state j if the process starts in state i. This is the *generalized renewal quantity*, a generalization of the renewal quantity of Chapter 3.

THEOREM 2.6 (Smith). Let $[P, \mathscr{F}(t)]$ be a finite semi-Markov process with $\mu_i < \infty$ for all i. Then

$$\lim_{t \to \infty} P_{ij}(t) = \frac{\mu_j}{l_{jj}}$$

with the understanding that t takes on only multiples of the span if G_{ij} is a lattice distribution, where $G_{ij}(t)$ denotes the distribution function for the first-passage time from state i to state j (see Smith, 1955, and Fabens, 1961).

We will need the following lemmas.

LEMMA 2.7.

$$\gamma_{jj}(t) = \sum_{k=1}^{\infty} G_{jj}^{(k)}(t)$$

and

$$\gamma_{ij}(t) = G_{ij}(t) + \int_0^t G_{ij}(t - x)\, d\gamma_{jj}(x)$$

where superscript k denotes k-fold convolution.

Proof. Let $N_{ij}(t)$ denote the number of occurrences of state j in $[0, t]$ if the process starts in state i. Then

$$P[N_{ij}(t) = 1] = G_{ij}(t) - \int_0^t G_{ij}(t - x)\, dG_{jj}(x),$$

and

$$P[N_{ij}(t) = n] = G_{ij} * G_{jj}^{(n-1)}(t) - G_{ij} * G_{jj}^{(n)}(t),$$

where the asterisk denotes convolution. Then

$$\gamma_{ij}(t) = \sum_{n=1}^{\infty} n[G_{ij} * G_{jj}^{(n-1)}(t) - G_{ij} * G_{jj}^{(n)}(t)]$$

$$= G_{ij}(t) + \int_0^t G_{ij}(t - x)\, d\gamma_{jj}(x). \;\|$$

LEMMA 2.8. If $G_{ij}(t)$ is absolutely continuous for all i and j, then

$P[\text{process enters state } j \text{ in } (t, t + dt) \,|\, \text{starts in } i]\, dt = d\gamma_{ij}(t)$.

Proof. The probability of the event in brackets is

$$dG_{ij}(t) + \sum_{n=1}^{\infty} dG_{ij} * G_{ij}^{(n)}(t) = d\gamma_{ij}(t). \;\|$$

THEOREM 2.9. If $[P, \mathscr{F}(t)]$ is a continuous-parameter semi-Markov process with each $l_{ij} < \infty$, then

$$\gamma_{jj}(t) = \frac{t}{l_{jj}} - 1 + \frac{1}{l_{jj}} \sum_k P_{jk}(t) l_{kj}.$$

Proof. We compute in two different ways the mean time until the first occurrence of state j after time t, if the process starts in state j. Let Y be

STOCHASTIC MODELS FOR COMPLEX SYSTEMS 137

a random variable denoting the time between successive occurrences of state j. Then

$$E[Y_1 + Y_2 + \cdots + Y_{N_{jj}(t)+1}] = l_{jj}[\gamma_{jj}(t) + 1]$$

by Theorem 2.4, Chapter 3. But this can also be written as

$$l_{jj}[\gamma_{jj}(t) + 1] = t + \sum_k P_{jk}(t) l_{kj}$$

by using the Markov character of the process. The result follows when we divide both sides by l_{jj}. ‖

THEOREM 2.10. If $[P, \mathscr{F}(t)]$ is a continuous-parameter semi-Markov process with $l_{ij} < \infty$, and $l_{jj}^{(2)} < \infty$, then

$$\gamma_{jj}(t) = \frac{t}{l_{jj}} + \frac{l_{jj}^{(2)}}{2l_{jj}^2} - \frac{l_{ij} + \tfrac{1}{2}w}{l_{jj}} + o(1)$$

where w is the period of $G_{jj}(t)$. This is an immediate consequence of Theorem 2.7 (extended) of Chapter 3.

THEOREM 2.11. If $[P, \mathscr{F}(t)]$ is a continuous-parameter Markov process, then

$$l_{jj}^{(2)} = 2l_{jj} \sum_{k=0}^{m} P_k{}^* l_{kj},$$

where $P_k{}^* = \lim\limits_{t \to \infty} P_{ik}(t)$.

Proof. From Theorem 2.10

$$\gamma_{jj}(t) = \frac{t}{l_{jj}} + \frac{l_{jj}^{(2)}}{2l_{jj}^2} - 1 + o(1)$$

and from Theorem 2.9

$$\gamma_{jj}(t) = \frac{t}{l_{jj}} - 1 + \frac{1}{l_{jj}} \sum_{k=0}^{m} P_k{}^* l_{kj} + o(1),$$

since $P_{ij}(t)$ tends to its limit exponentially fast for a finite-state Markov process. Hence

$$l_{jj}^{(2)} = 2l_{jj} \sum_{k=0}^{m} P_k{}^* l_{kj}. \;\|$$

These results have applications to Example 2.4 and to particular stochastic processes considered in the next section.

Example 2.4. In Example 1.1 states 4, 6, and 8 were designated unfavorable. The expected time spent in these three states in T hours is the mean down time in T hours. Mathematically, this is

$$\int_0^T [P_{05}(x) + P_{06}(x) + P_{08}(x)] \, dx$$

where $P_{ij}(x)$ is the probability that the process goes from state i to state j in time x. For large T,

$$\int_0^T P_{ij}(x)\, dx \sim \frac{\mu_j T}{l_{jj}}$$

where $l_{jj} = \dfrac{1}{\pi_j} \sum_{k=0}^{8} \pi_k \mu_k$; π_k $(k = 0, 1, \ldots, 8)$ were computed in Example 2.2, and μ_k $(k = 0, 1, \ldots, 8)$ were computed in Example 2.3. The error

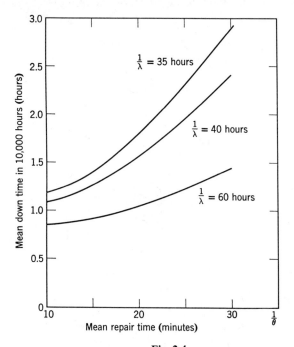

Fig. 2.4.

committed in the limit by using this approximation can be computed. The percentage error will be small for large T.

The mean number of total failures in $[0, T]$ is approximately

$$\frac{\pi_6 T}{\sum_{k=0}^{8} \pi_k \mu_k}.$$

The limiting probability of being in state 6, the state of total failure is

$$\lim_{x \to \infty} P_{06}(x) = \frac{\mu_6}{l_{66}}$$

by Theorem 2.6. These quantities are tabulated in Table 2.1 for several values of λ and θ. Figure 2.4 is a family of graphs of mean down time in hours for a 10,000-hour usage period. All computations are based on

$\gamma =$ time to perform preventive maintenance $= 1$ hour
$d =$ switchover time $= \frac{1}{6}$ hour
$t_0 =$ scheduled time until preventive maintenance $= 24$ hours.

TABLE 2.1

Mean Unit Life ($1/\lambda$ hr.)	Mean Repair Time ($1/\theta$ hr.)	Mean No. of Total Failures (in 10,000 hr.)	Mean Recurrence Time Between Total Failures	Mean Down Time (in 10,000 hr.)	$P_{06}(\infty)$
35	$\frac{1}{6}$	2.66	3,835	1.17	2.22×10^{-5}
35	$\frac{1}{2}$	7.84	1,293	2.90	1.96×10^{-4}
60	$\frac{1}{6}$	0.90	11,760	0.84	5.48×10^{-5}
60	$\frac{1}{2}$	2.69	3,771	1.43	6.73×10^{-5}

3. REPAIRMAN PROBLEMS

In this section we discuss a large class of commonly occurring stochastic models. In the literature these are known as repairman problems (Feller, 1957). These problems are related to queueing problems, and our main purpose is to exploit this relationship. For additional discussions of these problems see Cox and Smith (1961) and Takács (1962). In our discussion we follow an expository paper by Barlow (1962a).

Suppose we are given m identical units stochastically independent of one another and supported by n spare units. Unless otherwise stated, we assume all units are operating at $t = 0$. Suppose that each fails according to a distribution F. Furthermore, suppose that we have a repair facility capable of repairing s units simultaneously. Obviously, we could consider the facility as consisting of s repairmen. The following queue discipline is observed. If all repairmen are busy, each new failure joins a waiting line and waits until a repairman is freed. We assume that the repair times are also independent, identically distributed random variables with distribution G. The diagram in Figure 3.1 will be helpful in explaining various models of repairman problems.

One reason for studying repairman problems is to determine how reliability can be improved by using redundant units. In these cases it will be convenient to say that a total failure has occurred when all machines are being repaired or waiting to be repaired. We shall often be interested in the distribution of the time to the occurrence of a total failure and its moments.

When the failure distribution and the repair distribution are exponential, the repairman process just described generates a birth-and-death process. A very elegant treatment of this important class of processes has been given by Karlin and McGregor (1957). We will summarize their results of greatest interest in reliability problems.

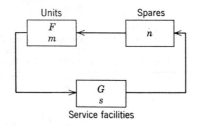

Fig. 3.1. Diagram illustrating repairman problems. From Barlow, Repairman problems, in *Studies in Applied Probability and Management Sciences*, edited by Arrow, Karlin, and Scarf, Stanford University Press, Stanford, Calif., 1962.

Birth-and-death processes

Unless otherwise indicated, we shall call the number of machines undergoing repair and waiting for repair the state of the process. When the failure and repair distributions are exponential, a birth-and-death process is generated. The states -1 and $N + 1 = m + n + 1$ are reflecting states. The reader will recall that a birth-and-death process is a stationary Markov process $X(t)$ whose state space is the nonnegative integers and whose transition probability matrix

$$P_{ij}(t) = P[X(t) = j \mid X(0) = i]$$

satisfies the conditions as $t \to 0$

$$P_{ij}(t) = \begin{cases} \lambda_i t + o(t), & j = i + 1 \\ \mu_i t + o(t), & j = i - 1 \\ 1 - (\lambda_i + \mu_i)t + o(t), & j = i \end{cases} \quad (3.1)$$

where $\lambda_i, \mu_i \geq 0$ for $i \geq 0$. The transition probability matrix $P(t) = [P_{ij}(t)]$ satisfies the differential equations

$$P'(t) = AP(t) \quad \text{and} \quad P'(t) = P(t)A,$$

with the initial condition $P(0) = I$, where A is a Jacobi matrix,

$$A = \begin{vmatrix} -\lambda_0 & \lambda_0 & 0 & 0 & \cdots & 0 & 0 \\ \mu_1 & -(\lambda_1 + \mu_1) & \lambda_1 & 0 & \cdots & 0 & 0 \\ 0 & \mu_2 & -(\lambda_2 + \mu_2) & \lambda_2 & \cdots & 0 & 0 \\ \cdot & \cdot & \cdot & \cdot \cdot & & \cdot & \cdot \\ \cdot & \cdot & \cdot & \cdot & \cdot & \cdot & \cdot \\ \cdot & \cdot & \cdot & \cdot & & \cdot & \cdot \\ 0 & 0 & 0 & 0 & \cdots & -(\lambda_{N-1} + \mu_{N-1}) & \lambda_{N-1} \\ 0 & 0 & 0 & 0 & \cdots & \mu_N & -\mu_N \end{vmatrix}$$

Associated with the matrix A is a system of polynomials $\{Q_k(x)\}$ defined by the recurrence formulas

$$Q_0(x) = 1,$$
$$-xQ_0(x) = -\lambda_0 Q_0(x) + \lambda_0 Q_1(x),$$
$$-xQ_k(x) = \mu_k Q_{k-1}(x) - (\lambda_k + \mu_k)Q_k(x) + \lambda_k Q_{k+1}(x), \qquad 0 < k \leq N$$

and normalized by the condition that $Q_k(0) \equiv 1$. The polynomial system $\{Q_k(x)\}$ is orthogonal with respect to a discrete measure ψ whose spectrum consists of the zeros of the polynomial

$$(\mu_N - x)Q_N(x) - \mu_N Q_{N-1}(x). \tag{3.2}$$

It is known that the zeros of these polynomials are all real and positive and that the zeros of $Q_k(x)$ and $Q_{k+1}(x)$ interlace. The polynomials are orthogonal in the sense that

$$\int_0^\infty Q_i(x)Q_j(x)\,d\psi(x) = \frac{\delta_{ij}}{\rho_j},$$

where $\rho_0 = 1$ and

$$\rho_j = \frac{\lambda_0 \lambda_1 \cdots \lambda_{j-1}}{\mu_1 \mu_2 \cdots \mu_j}.$$

A useful integral representation formula for these processes has been obtained by Karlin and McGregor (1957). They show that the transition probability $P_{ij}(t)$ can be represented by the formula

$$P_{ij}(t) = \rho_j \int_0^\infty e^{-xt} Q_i(x)Q_j(x)\,d\psi(x).$$

The measure ψ has a jump $1/\rho$, at $x = 0$, where $\rho = \sum_{j=0}^{N} \rho_j$. Hence the stationary probabilities are

$$P_j^* = \lim_{t \to \infty} P_{ij}(t) = \frac{\rho_j}{\rho}.$$

The Laplace transform of the first-passage distributions was obtained by Karlin and McGregor (1959a); that is,

$$G_{ij}^*(s) = \begin{array}{l} Q_i(-s)/Q_j(-s), \quad j > i \\ Q_{N-i}^*(-s)/Q_{N-j}^*(-s), \quad j < i \end{array}$$

where $\{Q_k^*(s)\}$ is the system of orthogonal polynomials corresponding to the same birth-and-death process, except that the states have been relabeled so that state N becomes the zero state, and so on. The corresponding parameters are

$$\lambda_k^* = \mu_{N-k} \quad \text{and} \quad \mu_k^* = \lambda_{N-k}.$$

Since
$$G_{0j}^*(s) = \frac{\lambda_0 \lambda_1 \cdots \lambda_{j-1}(-1)^j}{\prod_1^j (s + s_k)},$$

where $\{s_k\}_1^j$ are the roots of $Q_j(x)$, we obtain the following explicit formula for the density g_{0j} of G_{0j}:

$$g_{0j}(t) = (\lambda_0 \lambda_1 \cdots \lambda_{j-1})(-1)^j \sum_{k=1}^j \left[\frac{e^{-s_k t}}{\prod_{j \neq k}(-s_k + s_j)} \right]. \tag{3.3}$$

For birth-and-death processes this solves the general problem of determining the distribution of time until a total failure occurs.

Note that if Y_{ij} ($j > i$) denotes the first-passage time to go from i to j, then

$$Y_{ij} = Y_{0j} - Y_{0i}.$$

Hence $\quad l_{ij} = l_{0j} - l_{0i} \quad \text{and} \quad \sigma_{ij}^2 = \sigma_{0j}^2 - \sigma_{0i}^2.$

It is known that
$$l_{0j} = \sum_{k=0}^{j-1} \frac{1}{\lambda_k \rho_k} \sum_{r=0}^k \rho_r,$$

and that
$$\sigma_{0j}^2 = l_{0j}^2 - 2 \sum_{t=0}^{j-1} \frac{1}{\lambda_t \rho_t} \sum_{r=0}^t \rho_r l_{0r},$$

(Karlin and McGregor, 1959a).

The first moment of the recurrence time distribution is

$$l_{ii} = \frac{\rho}{\rho_i(\lambda_i + \mu_i)}.$$

By Theorem 2.11, we have

$$l_{ii}^{(2)} = \frac{2}{\rho_i(\lambda_i + \mu_i)} \sum_k \rho_k l_{ki}.$$

Using Theorem 2.9, we can obtain an exact formula for the expected number of total failures in $(0, t)$, namely,

$$\gamma_{NN}(t) = \frac{t\rho_N\mu_N}{\rho} - 1 + \frac{\rho_N\mu_N}{\rho} \sum_k P_{Nk}(t) l_{kN},$$

$$\gamma_{0N}(t) = \int_0^t [1 + \gamma_{NN}(t - x)] \, dG_{0N}(x).$$

(3.4)

Example 3.1. Consider the model shown in Figure 3.2. The symbol M stands for the exponential distribution. There are three states; state 0,

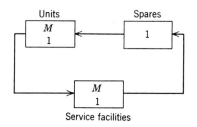

Fig. 3.2. From Barlow, Repairman problems, in *Studies in Applied Probability and Management Science*, edited by Arrow, Karlin, and Scarf, Stanford University Press, Stanford, Calif., 1962.

no machines down; state 1, exactly one machine down; state 2, exactly two machines down. The orthogonal polynomials are

$$Q_0(x) = 1,$$

$$Q_1(x) = \frac{\lambda - x}{\lambda},$$

$$Q_2(x) = \frac{1}{\lambda^2}[x^2 - (2\lambda + \mu)x + \lambda^2].$$

The zeros of $(\mu - x)Q_2(x) - \mu Q_1(x)$ are

$$x_0 = 0,$$
$$x_1 = (\lambda + \mu) + \sqrt{3\lambda\mu + \mu^2},$$
$$x_2 = (\lambda + \mu) - \sqrt{3\lambda\mu + \mu^2}.$$

Using (3.1) we can easily determine b_{ij} and c_{ij} where

$$P_{ij}(t) = \frac{\rho_j}{\rho} + b_{ij}e^{-x_1 t} + c_{ij}e^{-x_2 t}$$

since we know $P'_{ij}(0)$ and $P_{ij}(0)$.

The roots of $Q_2(x)$ are

$$s_1 = \frac{(2\lambda + \mu) + \sqrt{4\lambda\mu + \mu^2}}{2},$$

$$s_2 = \frac{(2\lambda + \mu) - \sqrt{4\lambda\mu + \mu^2}}{2}.$$

Therefore, using (3.3), we find that the density of $G_{02}(t)$ is

$$g_{02}(t) = \frac{\lambda^2 e^{-s_1 t}}{s_2 - s_1} - \frac{\lambda^2 e^{-s_2 t}}{s_2 - s_1}.$$

We can obtain $G_{20}(t)$ by simply interchanging the roles of λ and μ.

The embedded Markov chain of a birth-and-death process is a random walk. Many questions related to the repairman problem can be answered by considering this associated random walk. Let the random walk transition probabilities be p_i for $i \to i + 1$ and $q_i = 1 - p_i$ for $i \to i - 1$. Then

$$p_0 = 1,$$

$$p_i = \frac{\lambda_i}{\lambda_i + \mu_i}, \qquad 1 \leq i \leq N - 1$$

$$q_i = \frac{\mu_i}{\lambda_i + \mu_i}, \qquad 1 \leq i \leq N - 1$$

and $q_N = 1$. It is known (Harris 1952), that

$$\pi_j = \lim_{k \to \infty} p_{ij}^{2k} = \begin{cases} c, & j = 0 \\ c \dfrac{q_1 q_2 \cdots q_j}{p_0 p_1 \cdots p_{j-1}}, & j \neq 0 \end{cases}$$

where c is determined by the condition $\sum_{k=0}^{n} \pi_k = 1$. A quantity of some interest is the probability that the process, initially at i, reaches j at least once before returning to i. Call this probability θ_{ij}. A formula for θ_{ij} is

$$\theta_{ij} = \frac{p_i}{1 + q_{i+1}/p_{i+1} + \cdots + (q_{i+1} \cdots q_{j-1})/(p_{i+1} \cdots p_{j-1})}, \qquad i < j - 1$$

and $\theta_{j-1, j} = p_{j-1}$ (Harris, 1952).

The probability that the repairman process recovers or returns to the 0 state exactly r times before a total failure occurs is $(1 - \theta_{0N})^r \theta_{0N}$. The expected number of recoveries before a total failure is, of course, $(1 - \theta_{0N})/\theta_{0N}$. Similarly, the expected number of visits to state i, including the first before a total failure, is $1/\theta_{iN}$. Let T denote the expected total number of steps until a total failure. Then

$$T = \sum_{k=0}^{N} \frac{1}{\theta_{0k}},$$

and T includes one initial visit to each state. Each additional visit is caused by a repair. Hence $T + N$ is exactly twice the expected number of failures before a total failure.

Equal numbers of machines and repairmen—no spares

This model is identical with the continuous Ehrenfest model for diffusion processes. It was observed by Karlin and McGregor (1958) that the polynomial system is that of the Krawtchouk polynomials; that is,

$$Q_k(x) = \binom{N}{k}^{-1} \sum_{v=0}^{k} (-1)^v \binom{N-x}{k-v}\binom{x}{v}\left(\frac{\mu}{\lambda}\right)^v,$$

where $N = m$. The spectral measure ψ is the binomial distribution that places masses $\binom{N}{x} \lambda^x \mu^{N-x}$ at $x = 0, 1, \ldots, N$.

An explicit formula for the transition probabilities can be obtained for this process independently of the integral representation. Since there are as many repairmen as machines and since they are independent of one another, we may consider the machines and repairmen to be paired off. So really we have only N one-unit systems. By direct enumeration, we obtain

$$P_{ij}(t) = \sum_{k_1}\sum_{k_2} \binom{i}{k_1}\binom{N-i}{k_2}[p_{11}(t)]^{k_1}[p_{10}(t)]^{i-k_1}[p_{01}(t)]^{k_2}[p_{00}(t)]^{n-i-k_2},$$

where the summation extends over the set of integers defined by the relations $k_1 + k_2 = j$, $k_1 \leq i$, and $k_2 \leq N - i$, and

$$p_{00}(t) = \frac{\mu}{\lambda+\mu} + \frac{\lambda}{\lambda+\mu} e^{-(\lambda+\mu)t} \qquad p_{10}(t) = \frac{\mu}{\lambda+\mu} - \frac{\mu}{\lambda+\mu} e^{-(\lambda+\mu)t}$$

$$p_{01}(t) = \frac{\lambda}{\lambda+\mu} - \frac{\lambda}{\lambda+\mu} e^{-(\lambda+\mu)t} \qquad p_{11}(t) = \frac{\lambda}{\lambda+\mu} + \frac{\mu}{\lambda+\mu} e^{-(\lambda+\mu)t}.$$

It is easy to see that

$$P_j^* = \lim_{t\to\infty} P_{ij}(t) = \binom{N}{j}\left(\frac{\lambda}{\lambda+\mu}\right)^j\left(\frac{\mu}{\lambda+\mu}\right)^{N-j}.$$

Example 3.2. Consider the model shown in Figure 3.3. The orthogonal polynomials are

$$Q_0(x) = 1,$$

$$Q_1(x) = \frac{2\lambda - x}{2\lambda},$$

$$Q_2(x) = \frac{1}{2\lambda^2} x^2 - \frac{1}{2\lambda^2}(3\lambda + \mu)x + 1,$$

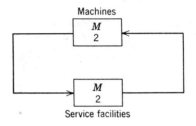

Fig. 3.3. From Barlow, Repairman problems, in *Studies in Applied Probability and Management Science*, edited by Arrow, Karlin, and Scarf, Stanford University Press, Stanford, Calif., 1962.

and the roots of $Q_2(x)$ are

$$s_1 = \frac{(3\lambda + \mu) + \sqrt{\lambda^2 + 6\lambda\mu + \mu^2}}{2}$$

$$s_2 = \frac{(3\lambda + \mu) - \sqrt{\lambda^2 + 6\lambda\mu + \mu^2}}{2}$$

and the density of the distribution of the time until a total failure occurs is

$$g_{02}(t) = \frac{2\lambda^2 e^{-s_1 t}}{s_2 - s_1} - \frac{2\lambda^2 e^{-s_2 t}}{s_2 - s_1}.$$

More general models

As we have seen for the exponential case, most questions of interest can be answered once we have obtained the associated orthogonal polynomials. Determining the roots of certain of these polynomials constitutes the principal difficulty. For nonexponential models, only special cases have been solved. The case of a general failure distribution, general repair distribution, one unit, one repairman, and no spares was discussed in Chapter 3, Section 6. We now consider two special cases. The first case contains an embedded semi-Markov process. The second case is remarkable in that any solution is possible. We define each by a diagram similar to Figure 3.1.

Case A. Solutions for this case were obtained by Takács (1957b). Define $\eta(t)$ as the number of machines working at the instant t. Let $\tau_1, \tau_2, \ldots, \tau_n, \ldots$ denote the end points of consecutive service times. Let $X_n = \eta(\tau_n^-)$. Then $\{X_n\}$ defines a Markov chain. Let $1/\lambda$ denote the

Case A

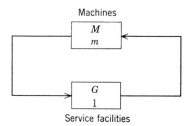

Fig. 3.4. From Barlow, Repairman problems, in *Studies in Applied Probability and Management Science*, edited by Arrow, Karlin, and Scarf, Stanford University Press, Stanford, Calif., 1962.

mean of the failure distribution; let β denote the mean of the repair distribution G, and let

$$\varphi(s) = \int_0^\infty e^{-sx}\, dG(x)$$

be the Laplace-Stieltjes transform of G. The stationary probabilities of $\{X_n\}$ were obtained in Takács (1957b). They are

$$\pi_k = \sum_{r=k}^{m-1} (-1)^{r-k} \binom{r}{k} B_r,$$

where B_r is the rth binomial moment of $\{\pi_k\}$, which is given by

$$B_r = \frac{C_r \sum\limits_{j=r}^{m-1} \binom{m-1}{j} \dfrac{1}{C_j}}{\sum\limits_{j=0}^{m-1} \binom{m-1}{j} \dfrac{1}{C_j}}$$

where $C_0 = 1$ and

$$C_r = \prod_{i=1}^{r} \frac{\varphi(i\lambda)}{1 - \varphi(i\lambda)}.$$

The limiting distribution

$$\lim_{t \to \infty} P[\eta(t) = i \mid \eta(0)] = P_i^*$$

was also obtained explicitly:

$$P_k^* = \frac{m\pi_{k-1}}{k(m\beta\lambda + \pi_{m-1})}, \qquad k = 1, 2, \ldots, m$$

and
$$P_0^* = 1 - \frac{m}{m\beta\lambda + \pi_{m-1}} \sum_{k=0}^{m} \frac{\pi_{k-1}}{k}.$$

The mean recurrence time to state 0 for this process is $1/\lambda P_1^*$. Similarly, the mean recurrence time to state m when all machines are working is

$$\frac{m\beta\lambda + \pi_{m-1}}{m\lambda\pi_{m-1}}.$$

The mean number of total breakdowns in $(0, t)$ is $\lambda t P_1^* + o(t)$. Takács also obtains many other results of interest for the nonstationary process.

Case B. This model can be identified with the telephone trunking problem with Poisson input to a finite number of channels. Note that

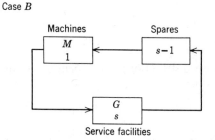

Fig. 3.5. From Barlow, Repairman problems, in *Studies in Applied Probability and Management Science*, edited by Arrow, Karlin, and Scarf, Stanford University Press, Stanford, Calif., 1962.

when all service facilities are busy, no machine is operating. Since an exponential distribution is invariant under truncation on the left, we see that this repairman problem is the same as the telephone trunking model with Poisson input. The stationary probabilities for the number of machines down are expressed by Erlang's formula which follows. Let $Y(t)$ denote the state of the process (i.e., the number of repair facilities busy at time t). Let $1/\lambda$ denote the mean of the failure distribution and β the mean of the repair distribution G. It is shown in Sevastjanov (1957) that

$$P_k^* = \lim_{t \to \infty} P[Y(t) = k \,|\, Y(0)] = \frac{(\lambda\beta)^k/k!}{\sum_{i=0}^{s}(\lambda\beta)^i/i!}.$$

Since the state 0 is a regeneration point or renewal point, we can obtain the mean recurrence time to state 0 by applying the fundamental renewal theorem (Theorem 2.9, Chapter 3)

$$P_0(t) = e^{-\lambda t} + \int_0^t e^{-\lambda x} \, dM_0(t - x),$$

$$\lim_{t \to \infty} P_0(t) = \frac{1/\lambda}{l_{00}}, \quad l_{00} = \frac{1}{\lambda P_0^*}.$$

General failure and exponential-repair distributions

We consider two cases closely allied to those already discussed. Let the failure distribution be F with mean α. Let the mean time for repair be $1/\mu$.

Case C

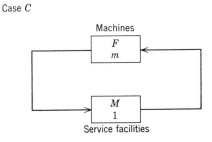

Fig. 3.6. From Barlow, Repairman problems, in *Studies in Applied Probability and Management Science*, edited by Arrow, Karlin, and Scarf, Stanford University Press, Stanford, Calif., 1962.

Case C. This has the same stochastic structure as Case B. The stationary probability P_k^* that k machines are operating is

$$P_k^* = \frac{(\mu\alpha)^k/k!}{\sum_{j=0}^m (\mu\alpha)^j/j!}.$$

The mean recurrence time to the state of total failure when no machines are operating is $1/\mu P_0^*$.

Case D. This yields the same limiting process as in Case A. The stationary probabilities (P_k^*) listed there become the stationary probabilities for the number of machines undergoing repair in this model. We replace λ by μ, m by s, and β by α. The mean recurrence time to state 0, when all machines are operative, is $1/\mu P_1^*$. Similarly, the mean recurrence time to the state s, when all machines are being repaired, is

$$\frac{s\alpha\mu + \pi_{m-1}}{s\mu\pi_{m-1}}.$$

Case *D*

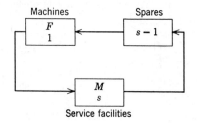

Fig. 3.7. From Barlow, Repairman problems, in *Studies in Applied Probability and Management Science*, edited by Arrow, Karlin, and Scarf, Stanford University Press, Stanford, Calif., 1962.

Remark

More general models, such as those shown in Figure 3.8, can also be solved by using the technique of the embedded semi-Markov process. To illustrate the first case, let $\tau_1, \tau_2, \ldots$ denote the end points of consecutive service periods. Let $\eta(t)$ denote the number of machines operating and operative at time t. The embedded Markov chain is defined by

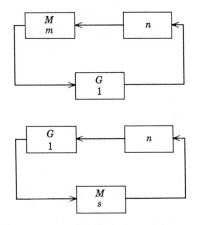

Fig. 3.8. From Barlow, Repairman problems, in *Studies in Applied Probability and Management Science*, edited by Arrow, Karlin, and Scarf, Stanford University Press, Stanford, Calif., 1962.

$X_n = \eta(\tau_n^-)$. The stationary probabilities for this case were obtained by Takács (1958), who solved a telephone trunking problem that generates the same embedded Markov chain as that which results if we let $w = n - 1$, where w is the maximum permissible waiting line for calls. His formulas, though usable, are exceedingly complicated and will not be reproduced here. The stationary probabilities for the actual number of machines working can be obtained in a straightforward fashion from the stationary probabilities of the Markov chain.

TABLE 3.1

STOCHASTIC PROCESSES ASSOCIATED WITH REPAIRMAN PROBLEMS*

Failure Dist.	Repair Dist.	No. Machines	No. Repairmen	No. Spares	Process
M	M	m	s	n	B and D
M	M	1	s	n	B and D Truncated queueing
M	M	m	1	n	B and D Truncated queueing
M	M	m	m	0	B and D Ehrenfest model
G	M	1	s	$s-1$	SMP
M	G	m	1	0	SMP
G	M	m	1	0	Erlang's formula
M	G	1	s	$s-1$	Erlang's formula
G	G	1	1	0	SMP
G	M	1	s	n	SMP
M	G	m	1	n	SMP

B and D = Birth-and-death processes; SMP = Semi-Markov process.

* From Barlow, Repairman problems, in *Studies in Applied Probability and Management Science*, edited by Arrow, Karlin, and Scarf, Stanford University Press, Stanford, Calif., 1962.

4. MARGINAL CHECKING

In Chapter 4 we obtained optimal replacement policies assuming the unit to be replaced could be in only two states—an operative and a non-operative state. In this section and the next we suppose that our unit or system is capable of assuming many states. For example, in a circuit described by r component parameters, the state of a circuit is defined as an r-tuple, each element of which denotes a state of one of the component parameters. Thus, in a circuit consisting of r parameters in which the range of values of each parameter is partitioned into N distinct states, we have N^r circuit states. The circuit parameters will have a tendency to drift out of specified tolerance bounds. It is the purpose of a marginal test to determine the state of the parameters and on the basis of a marginal checking policy to decide whether or not to replace the circuit.

This type of periodic checking of a unit is a common industrial and military procedure. Given that the checking interval and the ways of modifying the unit are fixed, an important problem is that of determining, according to some reliability criterion, the optimal policy for making the appropriate decision. In this section we assume that the only possible altering of the unit consists in replacing it.

Drenick (1960a) considered the problem of choosing a decision rule for replacing a unit based on a marginal test and of determining an optimal checking interval. Flehinger (1962b) considered a model where it is assumed that a unit may be in one of $m + 1$ states $0, 1, 2, \ldots, m$ and, during normal operation, this is the state space of a continuous-parameter Markov process in which state m is the failed state. When a unit enters state m, it is immediately replaced by one in state 0. The marginal test detects the state of the unit, and states $k, k + 1, \ldots, m - 1$ are considered marginal. The test is performed at fixed intervals, and, if a unit is found in the marginal state, it is replaced by one in state 0. Flehinger determined the operating characteristics of policies which we call *control limit rules* (Derman, 1963c). A control limit rule is of the simple form: *Replace the unit if and only if the observed state is one of the states* $k, k + 1, \ldots, m$ *for some k*. The set of states $k, k + 1, \ldots, m - 1$ will be called marginal states. The state k is called the control limit. The marginal testing problem is very conveniently viewed as a discrete one by assuming that failures can occur only at the end of a checking interval. If the checking interval is specified, the only problem is to determine the set of states which we call marginal.

Formal model

The following formulation is due to Derman (1963c). Suppose that a unit is inspected at equally spaced points in time and that after each inspection it is classified into one of $m + 1$ states $0, 1, 2, \ldots, m$. A unit is in state 0 (m) if and only if it is new (inoperative). Let the times of inspection be $n = 0, 1, \ldots$ and let $X(n)$ denote the observed state of the unit in use at time n. We assume that $\{X(n)\}$ is a Markov chain with stationary transition probabilities

$$q_{ij} = P[X(n + 1) = j \mid X(n) = i]$$

for all i, j, and n. The actual values of the q_{ij}'s are functions of the nature of the unit and the original replacement rule: *Replace only when the unit is inoperative*. Hence we assume

$$q_{j0} = 0, \quad \text{for } j < m$$
$$q_{jm}^n > 0, \quad \text{for some } n \geq 1 \text{ and each } j < m$$

and $\quad q_{m0} = 1.$

We suppose a cost of c_1 ($c_1 > 0$) is incurred if the unit is replaced after becoming inoperative, a cost c_2 ($0 < c_2 < c_1$) if the unit is replaced before it becomes inoperative; otherwise, no cost is incurred. The criterion for comparing replacement rules is the average cost per unit time averaged over a large (infinite) time interval.

Derman (1962) has shown that we may confine attention to nonrandomized stationary rules; that is, the optimal replacement rule will be a partitioning of the state space into two categories: marginal states at which the unit is replaced and states at which it is not replaced. Such a replacement rule will result in modifying the original Markov chain. Let (p_{ij}) denote the transition matrix of the modified chain. Then $p_{j0} = 1$ for one or more (but not all) of the j's for $0 < j < m$. Such a modification corresponds, in the replacement context, to replacing the item if it is observed to be in state j. Since there are $2^{m-1} - 1$ such possible modifications, each corresponding to a possible partitioning of the state space, an optimal modification exists. For small values of m, the optimal modification is selected by enumerating and computing. In the next section, we provide an algorithm based on linear programming for computing the optimal solution.

Let $\mathscr{C}$ denote the class of possible replacement rules just described. For each rule R in $\mathscr{C}$ let (p_{ij}) denote the associated set of transition probabilities and consider the cost function

$$c(j) = \begin{cases} 0, & \text{if } p_{j0} = 0, j < m \\ c_2, & \text{if } p_{j0} = 1, j < m \\ c_1, & \text{if } j = m. \end{cases}$$

Thus $c(j)$ denotes the cost incurred at any time n when the Markov chain is in state j. Let $A(R)$ denote the asymptotic average expected cost if we follow rule R. Then

$$A(R) = \lim_{n \to \infty} \frac{1}{n} \sum_{k=1}^{n} c[X(k)] = \sum_{j=0}^{m} \pi_j c(j) \qquad (4.1)$$

with probability one, where the steady-state probabilities π_j satisfy the equations

$$\pi_j = \sum_{i=0}^{m} \pi_i p_{ij}, \quad j = 0, 1, \ldots, m$$

$$\sum_{j=0}^{m} \pi_j = 1,$$

and
$$0 \leq \pi_j \leq 1, \quad j = 0, 1, \ldots, m.$$

(Chung, 1961, p. 87).

IFR Markov chains

The main result of this section is that the optimal replacement rule is a control limit rule if the underlying Markov chain with transition probability matrix (q_{ij}) is IFR.

Definition. A Markov chain is said to be IFR if

$$P[X(n + 1) \in B \mid X(n) = i] \qquad (4.2)$$

is nondecreasing in i for every set B of the form $B = \{k, k + 1, \ldots, m\}$ for some $k = 0, 1, \ldots, m$; that is,

$$h(i) = \sum_{j=k}^{m} q_{ij} \qquad (4.3)$$

is nondecreasing in i for all $k = 0, 1, \ldots, m$.

The condition is quite intuitive if "the greater the unit deterioration, the greater the unit state value." Thus for every "final" set B, the larger the value of i, the more probable it is that the unit will enter the "final" set at the next inspection. Note that (4.2) is a conditional probability analogous to the failure rate function discussed in Chapter 2.

A stronger condition than IFR and one which is easy to check can be given in terms of the transition matrix (q_{ij}). If (q_{ij}) is TP$_2$ in $i = 0, 1, \ldots, m - 1$ and $j = 0, 1, \ldots, m$ (see Appendix 1), then the Markov chain is IFR. From the variation diminishing property of TP$_2$ functions (Appendix 1) we see that (4.3) must be nondecreasing in $i = 0, 1, \ldots, m - 1$ and hence in $i = 0, 1, \ldots, m$ since $h(m) = 1$ if $k = 0$ and $h(m) = 0$ otherwise.

A common class of continuous time parameter stochastic processes describing unit operation are the birth-and-death processes discussed in Section 3. If $P_{ij}(t)$ denotes the transition probability for such a process and

$$q_{ij} = \begin{cases} P_{ij}(t), & \text{if } i \neq m \\ 0, & \text{if } i = m, j \neq 0 \\ 1, & \text{if } i = m, j = 0 \end{cases}$$

for fixed $t > 0$, then (q_{ij}) is IFR. This follows from the total positivity of $P_{ij}(t)$ in i and j for fixed $t > 0$ (Karlin and McGregor, 1957). Note, however, that (q_{ij}) is *not* TP$_2$ in $i = 0, 1, \ldots, m$ and $j = 0, 1, \ldots, m$ since $q_{m,0} = 1$. Hence the Markov chain which arises by inspecting a birth-and-death process at times $t, 2t, 3t, \ldots$ is IFR.

The following theorem is proved by Derman (1963c).

THEOREM 4.1 (Derman). If the Markov chain with transition probabilities (q_{ij}) is IFR, then there exists a control limit rule R^* such that

$$A(R^*) = \min_{R \in \mathscr{C}} A(R).$$

Optimal marginal checking intervals

So far we have confined attention to the problem of determining the set of states to be designated marginal. Often the set of marginal states are specified and we desire an optimal set of checking intervals.

Suppose now that a given unit (system, etc.) can be described operationally by a pure death process; that is, the process is a continuous-parameter Markov process with $m + 1$ states $0, 1, \ldots, m$ such that if the process is in state i, the next state to be entered is $i + 1$ and state m is an absorbing (failed) state. For example, consider a unit composed of m identical components in parallel. Suppose all components fail independently and exponentially with parameter λ. Let the number of failed components, say i, at time t denote the state of the unit. Then

$$P_{ij}(t) = \binom{m-i}{j-i}(1 - e^{-\lambda t})^{j-i} e^{-(m-j)\lambda t}$$

for $i \leq j \leq m$ denotes the probability that the unit proceeds from state i to state $j \geq i$ in time t.

In general, we shall call the states $k, k+1, \ldots, m-1$ marginal and suppose that a cost c_1 is incurred for replacing a failed unit and a cost $c_2 < c_1$ is incurred for replacing the unit if it is in the marginal set. Let c_3 denote the cost of inspecting the unit irrespective of whether a replacement is subsequently made. Assuming that the state of the unit is discovered only by checking, we desire a sequence of numbers $x_0, x_1, \ldots, x_{k-1}$ such that if the process is observed to be in state i, the next inspection is scheduled x_i hours later. Let $\mathscr{L}_i(x)$ denote the expected one-period loss assuming that the unit is in state i $(0 \leq i < k)$, that we inspect the unit x hours later, and that we proceed in an optimal fashion thereafter. Clearly, the optimal checking sequence can be determined recursively. Then $\mathscr{L}_{k-1}(x)$ satisfies the following equation:

$$\mathscr{L}_{k-1}(x) = c_2 \sum_{j=k}^{m-1} P_{k-1,j}(x) + c_1 P_{k-1,m}(x) + P_{k-1,k-1}(x)\mathscr{L}_{k-1}(x) + c_3.$$

Solving for $\mathscr{L}_{k-1}(x)$ we let x_{k-1} satisfy

$$\mathscr{L}_{k-1}(x_{k-1}) = \min_x \frac{c_2 \sum_{j=k}^{m-1} P_{k-1,j}(x) + c_1 P_{k-1,m}(x) + c_3}{1 - P_{k-1,k-1}(x)}$$

$$= \min_x \left[c_2 + (c_1 - c_2)\frac{P_{k-1,m}(x) + c_3}{1 - P_{k-1,k-1}(x)} \right].$$

and

$$\mathscr{L}_{k-2}(x_{k-2}) = \min_x \frac{c_2 \sum_{j=k}^{m-1} P_{k-2,j}(x) + c_1 P_{k-2,m}(x) + P_{k-2,k-1}(x)\mathscr{L}_{k-1}(x_{k-1}) + c_3}{1 - P_{k-2,k-2}(x)}$$

and so on.

By computing the optimal checking sequence and comparing values of $\mathscr{L}_0(x_0)$ for $k = 0, 1, 2, \ldots, m$, we can optimize over both k and the checking sequence.

We have touched on only a few of many possible models involving marginal checking. The techniques discussed here should be useful in solving many related problems, however.

5. OPTIMAL MAINTENANCE POLICIES UNDER MARKOVIAN DETERIORATION

We consider a system with a finite number $m + 1$ of states $i = 0, 1, 2, \ldots, m$. Periodically, say once a day, we observe the current state of the system, and then choose some decision procedure d_k ($k = 1, 2, \ldots, K$) with a certain probability depending on the past history of the system. As a joint result of the current state i of the system and the decision procedure d_k chosen, two things happen: (1) we suffer an expected cost c_{ik} and (2) the system moves to a new state j with the probability of a particular new state j given by $q_{ij}(k)$. If there is specified a discount factor β ($0 \leq \beta < 1$), so that the present value of unit income n days in the future is β^n, our problem is to choose a policy which minimizes our total expected loss. We shall be more interested in minimizing the average expected cost per unit time whenever the limit exists, however.

The model with discounting has been studied by Howard (1960) and Blackwell (1962). The model with average expected cost per unit time has been treated by Derman (1962). These authors prove the existence of an optimal policy. Derman reduces the problem to a problem in linear programming following the ideas of Manne (1960). Jewell (1963) has extended these results to Markov renewal processes. The following formulation is due to Derman (1962). Let $X_0, X_1, \ldots$ denote the sequence of observed states of the system and $\Delta_0, \Delta_1, \ldots$ the sequence of decisions. Derman (1963a) has shown that under suitable restrictions we may confine attention to so-called randomized stable rules; that is,

$$P[\Delta_t = d_k \mid X_0, X_1, \ldots, X_t = i] = D_{ik}$$

so that our decision rule depends only on the last state observed. In fact, the optimal decision rule will be a stable one, that is, $D_{ik} = 0$ or 1. For our linear programming formulation, however, it will be convenient to consider the larger class of randomized stable rules. We shall assume throughout that

$$\sum_{k=1}^{K} D_{ik} = 1$$

$$P[X_{t+1} = j \mid X_0, \ldots, X_t = i] = \sum_{k=1}^{K} q_{ij}(k) D_{ik}$$

for $i, j = 0, 1, \ldots, m$; $t = 0, 1, \ldots$ where $(q_{ij}(k))$ are such that

$$q_{ij}(k) \geq 0, \quad i, j = 0, 1, \ldots, m; \; k = 1, \ldots, K.$$

Given that the system is in state i, $q_{ij}(k)$ is the probability that it next appears in state j when we employ policy k.

Let

$$p_{ij} = \sum_{k=1}^{K} q_{ij}(k) D_{ik}, \quad i, j = 0, 1, \ldots, m$$

and note that (p_{ij}) is a stationary transition matrix of a Markov chain. We shall assume that for every randomized decision rule employed, the states $0, 1, \ldots, m$ of the resulting Markov chain belong to the same ergodic class.

Let $c_{ik}(t) \geq 0$ $(i = 0, 1, \ldots, m; \; k = 1, \ldots, K; \; t = 0, 1, \ldots)$ be finite values denoting the expected cost incurred at time t given that the system is observed in state i at time t and that decision d_k is made. We assume that $c_{ik}(t) = c_{ik}$, independent of t.

Let C_t $(t = 0, 1, \ldots)$ denote the expected cost incurred at time t as a result of using a given decision procedure R. We want to obtain that decision procedure R which minimizes

$$\limsup_{T \to \infty} \frac{1}{T} \sum_{t=0}^{T} C_t,$$

the average expected cost per unit time, whenever the limit exists.

The following results from Markov chain theory will be relevant (see Section 2.1). Let (p_{ij}) be the transition probabilities of a Markov chain $\{X_t\}$ $(t = 0, 1, \ldots)$ with a finite set I of states all belonging to the same class; let $f(j)$ $(j \in I)$ be a function defined over the states. Then

$$\lim_{T \to \infty} \frac{1}{T} \sum_{t=0}^{T} E[f(X_t)] = \sum_{j \in I} \pi_j f(j) \tag{5.1}$$

[see (4.1)] where the π_j uniquely satisfy

$$\pi_j \geq 0$$
$$\pi_j - \sum_{i \in I} \pi_i p_{ij} = 0, \quad j \in I \tag{5.2}$$
$$\sum_{j \in I} \pi_j = 1.$$

In fact, $\pi_j > 0$ for all $j \in I$.

We shall show, assuming that all states belong to the same class for every randomized decision rule, that the problem of obtaining an optimal decision rule can be formulated as a linear programming problem.

Setting

$$f(j) = \sum_{k=1}^{K} D_{jk} c_{jk}, \qquad j = 0, 1, \ldots, m$$

we have from (5.1) that for every randomized decision rule R,

$$\lim_{T \to \infty} \frac{1}{T} \sum_{t=0}^{T} C_t = \sum_{j=0}^{m} \pi_j \sum_{k=1}^{K} D_{jk} c_{jk}, \qquad i = 0, 1, \ldots, m. \qquad (5.3)$$

Let $x_{jk} = \pi_j D_{jk}$ ($j \in I$; $k = 1, \ldots, K$.) Intuitively, x_{jk} is the stationary probability of making decision k when the system is in state j. Then (5.3) becomes

$$\sum_{j=0}^{m} \sum_{k=1}^{K} x_{jk} c_{jk},$$

and (5.2) becomes

$$x_{jk} \geq 0, \qquad j \in I, k = 1, \ldots, K$$

$$\sum_{k=1}^{K} x_{jk} - \sum_{i \in I} \sum_{k=1}^{K} x_{ik} q_{ij}(k) = 0, \qquad j \in I \qquad (5.4)$$

$$\sum_{j \in I} \sum_{k=1}^{K} x_{jk} = 1.$$

Because of the uniqueness of the solutions to (5.2), there corresponds to every randomized decision rule a solution to (5.4) with $\sum_{k=1}^{K} x_{jk} > 0$, $j \in I$. Moreover, every solution to (5.4) must satisfy $\sum_{k=1}^{K} x_{jk} > 0$, $j \in I$; hence, setting

$$D_{jk} = \frac{x_{jk}}{\sum_{k=1}^{K} x_{jk}}, \qquad j \in I; k = 1, \ldots, K \qquad (5.5)$$

a solution to (5.4) corresponds to some randomized decision rule.

Thus, the solution to our problem consists in minimizing

$$\sum_{j=0}^{m} \sum_{k=1}^{K} x_{jk} c_{jk}$$

subject to the constraints (5.4). We refer the reader to Dantzig (1963) for methods to solve linear programming problems.

Optimal maintenance policies with probability bounds on failure

Derman (1963b) has modified the foregoing loss structure to solve a replacement problem of some practical importance. We assume the system can be in one of $m + 1$ states, labeled $0, 1, \ldots, m$, as before. However, we now identify state 0 with a new system, states $m - r + 1, \ldots, m$, as inoperative states from which replacement must follow, and states $1, \ldots, m - r$ as operative states from which it is possible to replace the system or maintain it according to one of the various maintenance procedures labeled as d_k $(k = 1, \ldots, K)$.

As before,

$$q_{ij}(k) = P[X_{t+1} = j \mid X_0, \Delta_0, \ldots, X_t = i, \Delta_t = d_k],$$
$$t = 0, 1, \ldots; i, j = 0, \ldots, m; k = 1, \ldots, K,$$

where the $q_{ij}(k)$'s are nonnegative numbers satisfying

$$\sum_{j=1}^{m} q_{ij}(k) = 1, \quad i = 0, 1, \ldots, m; k = 1, \ldots, K.$$

The $q_{ij}(k)$'s are assumed known. If the system is in state i, and the decision rule d_k followed is to replace the system, then $q_{i0}(k) = 1$. We also assume that $q_{0j}(k) > 0$; $j = 1, \ldots, m$; $k = 1, \ldots, K$, that is, that the system can go to any state in one time period, and that the system, if not replaced, will eventually reach one of the inoperative states $m - r + 1, \ldots, m$.

Consider now the situation in which no precise cost structure is given, merely that the costs of replacement are of higher order of magnitude than the costs of maintenance and that the costs of being inoperative are of higher order of magnitude than the costs of replacement. Thus, in the absence of specific cost information we consider the problem of maximizing the expected length of time between replacements, subject to the constraints that, in any one cycle, the probability of replacement via state j $(j = m - r + 1, \ldots, m)$ is no greater than a preassigned value a_j. The numbers a_j $(j = m - r + 1, \ldots, m)$ should be such that they reflect the undesirability of a forced replacement through state j; if such quantitative assessment is impracticable, particularly when such forced replacement is catastrophic, the a_j can be made sufficiently small to give the kind of protection usually deemed "safe."

Assume $X_0 = 0$. Let η denote the smallest positive integer t such that $X_t = 0$, and let τ_j be the smallest positive integer t such that $X_t = j$ $(j = m - r + 1, \ldots, m)$. The problem is to choose the rule R, over all possible rules, such that $E\eta$ is maximized, subject to the constraints $P(\tau_j < \eta) \leq a_j$ $(j = m - r + 1, \ldots, m)$.

Note that $E\eta$ and $P(\tau_j < \eta)$ involve only those X_t's for which $t \leq \eta$. Thus we can restrict ourselves to considering rules R for which state 0 is a recurrent event; that is, whenever the system enters state 0, subsequent decisions are made as if the system were starting from $t = 0$. We shall refer to such rules as cyclic rules; a new cycle begins at each entry into state 0.

It can be shown, using the strong law of large numbers, that for cyclic rules

$$\lim_{T\to\infty} \sum_{t=1}^{T} P(X_t = j \mid X_0 = 0) = \frac{\theta_{0j}}{E\eta}, \quad j = 0, \ldots, m \quad (5.6)$$

where θ_{0j} denotes the expected number of times j occurs within one cycle. By the way in which we have defined a cycle, $\theta_{00} = 1$; therefore, from (5.6) follows

$$E\eta = 1/\pi_{00} \quad (5.7)$$

where π_{0j} is the left member of (5.6). Moreover, since the states j ($j = m - r + 1, \ldots, m$), are entered at most once in any cycle, and only when $\tau_j < \eta$,

$$\theta_{0j} = P(\tau_j < \eta), \quad j = m - r + 1, \ldots, m. \quad (5.8)$$

Hence, from (5.6), (5.7), and (5.8) we have

$$P(\tau_j < \eta) = \pi_{0j}/\pi_{00}, \quad j = m - r + 1, \ldots, m. \quad (5.9)$$

The problem under consideration, restated, is that of finding the rule R, in the class of all cyclic rules, that minimizes π_{00} subject to the constraints $\pi_{0j} - a_j \pi_{00} \leq 0$ ($j = m - r + 1, \ldots, m$).

The randomized stable rules form a subclass of the cyclic rules. For any randomized stable rule R, the sequence $\{X_t\}$ ($t = 0, 1, \ldots$) is an irreducible Markov chain with stationary transition probabilities

$$p_{ij} = \sum_{k=1}^{K} q_{ij}(k) D_{ik}, \quad i, j = 0, \ldots, m. \quad (5.10)$$

The fact that it is irreducible follows from the assumptions concerning the $q_{ij}(k)$. Furthermore, from Markov chain theory, $\pi_{0j} = \pi_j$ ($j = 0, \ldots, m$), where the π_j are positive numbers uniquely satisfying

$$\sum_{i=0}^{K} \pi_i p_{ij} = \pi_j \geq 0, \quad j = 0, \ldots, m$$

$$\sum_{j=0}^{m} \pi_j = 1. \quad (5.11)$$

A consequence of the theorem proved by Derman (1963a) is that, in seeking the optimal rule in question, it is sufficient to consider only the class of randomized stable rules. Therefore the problem can be restated again: Choose the D_{ik}'s such that π_0 is minimized, subject to the constraint that (5.11) is satisfied and that

$$\pi_j - a_j \pi_0 \leq 0 \quad (j = m - r + 1, \ldots, m).$$

The method used in the preceding problem now applies. Let $x_{ik} = \pi_i D_{ik} \geq 0$, $(i = 0, \ldots, m; k = 1, \ldots, K)$. Then $\pi_i = \sum_{k=1}^{K} x_{ik} > 0$, $(i = 0, \ldots, m)$; $D_{ik} = x_{ik} \Big/ \sum_{k=1}^{K} x_{ik}$ $(i = 0, \ldots, m; k = 1, \ldots, K)$.

The equations (5.11) now become, using (5.10),

$$\sum_{i=0}^{m} \sum_{k=1}^{K} x_{ik} q_{ij}(k) = \sum_{k=1}^{K} x_{jk}, \quad j = 0, \ldots, m$$

$$\sum_{i=0}^{m} \sum_{k=1}^{K} x_{ik} = 1. \tag{5.12}$$

Consequently, the problem can be stated in the linear programming form. Choose the x_{ik}'s such that $\sum_{k=1}^{K} x_{0k}$ is minimized, subject to the constraints that $x_{ik} \geq 0$, that (5.12) holds, and that

$$\sum_{k=1}^{K} x_{jk} - a_j \sum_{k=1}^{K} x_{0k} \leq 0, \quad j = m - r + 1, \ldots, m.$$

It is clear that a feasible solution will exist if and only if for some k, $q_{0j}(k) \leq a_j$ $(j = m - r + 1, \ldots, m)$.

CHAPTER 6

Redundancy Optimization

1. INTRODUCTION

In this chapter we present models concerned with optimization of redundancy. The first problem treated, that of maximizing the reliability of a series system subject to one or more constraints on total cost, weight, volume, etc., is one of the early important problems of reliability. The problem has a number of variations depending on whether redundancy is parallel (redundant units operating simultaneously in parallel) or standby (redundant units standing by as spares and used successively for replacement), and also depending on whether or not specific values for the constraints are given in advance. In Section 2 the basic variations of the problem are explicitly stated.

For the one-constraint parallel redundancy case (see models A and B of Section 2 of this chapter), Mine (1959) and Moskowitz and McLean (1956) obtain approximate solutions. Kettelle (1962) provides an algorithm based on dynamic programming for obtaining an exact solution. For the two-constraint parallel redundancy case, Bellman and Dreyfus (1958) sketch a dynamic programming method for maximizing system reliability given specific constraint values. For the standby redundancy case (see models C and D of Section 2), Geisler and Karr (1956) and Gourary (1956, 1958) minimize the expected value of shortage weighted by the essentiality of the item short, subject to a single specific constraint. The objective function they optimize is thus different from the one we consider, namely, system survival probability. For discussions of allocation problems in general, see Everett (1963) and Zahl (1963).

Many papers (including the references just mentioned) have treated the problem of achieving optimum redundancy assuming only two component states are possible, the operating state and the failed state. In some situations, however, it would be more realistic to further subdivide failure into two classes depending on the effect of the failure on system operation. For example, in a network of relays in series, an open-circuit failure of a single relay will render the system unresponsive, since a signal cannot complete a path through the network; on the other hand, short-circuit failure of *all* the relays will also render the system unresponsive since in this case a signal will always complete a path through the network whether this is desired or not. Insertion of additional components in series increases the chance of system failure caused by open-circuit failure of a single relay at the same time that it decreases the chance of system failure caused by short-circuit failure of all the relays in the network. Thus a valid problem exists in determining the optimum number of relays to arrange in series. No constraints are considered in this problem.

For relay networks Moore and Shannon (1956) show how to design systems so as to achieve arbitrarily high system reliability given relays subject to open- and short-circuit failure with specified probabilities (see Section 2 of Chapter 7). Hanne (1962) considers circuits consisting of up to four components, each subject to open- and short-circuit failures; he obtains system reliability and mean system life assuming the exponential distribution for component life. Barlow, Hunter, and Proschan (1963) show how to determine the number of series subsystems in parallel to maximize system reliability. They also obtain qualitative relationships between this optimum number and the various parameters present. Section 6 of this chapter is based on this analysis. Gordon (1957) treats a similar model but maximizes a somewhat different objective function.

2. OPTIMAL ALLOCATION OF REDUNDANCY SUBJECT TO CONSTRAINTS

Introduction

A basic dilemma exists in incorporating redundancy into the design of systems so as to assure high reliability. On the one hand, we should like to provide for each vital component of the system as many redundant units as possible. On the other hand, we do not wish to design an excessively costly, heavy, or bulky system. In fact, there may be specific constraints on cost, weight, or volume, or even on several such factors simultaneously. How shall we achieve an optimal allocation of redundancy, that is, maximum system reliability for the cost, weight, or volume, etc., allowed?

We shall confine ourselves to series-type systems; that is, we assume the system, consisting of k stages (or subsystems), functions if and only if each stage functions. Even within this limitation, however, a number of distinct variations of the basic problem exist.

1. Redundant units may be operating actively in parallel and thus be subject to failure. Alternately, they may be serving as spares to be used in succession for replacement of failed units; in this case redundant units are not subject to failure while in standby condition. The first type of redundancy is sometimes referred to as parallel redundancy, the second as standby redundancy.

2. Fixed specific constraints may exist on cost, weight, volume, etc., which cannot be violated. What is desired in this case is a redundancy allocation which satisfies these constraints while maximizing reliability. Alternately, fixed constraint values may not be available for cost, weight, volume, etc. Rather, what may be desired is a family of redundancy allocations, each member of the family having the optimality property that any redundancy allocation achieving higher reliability must be either costlier, heavier, or bulkier. Such a family of optimal redundancy allocations, although still containing many members, represents a tremendous reduction of redundancy allocations that the decision maker needs to consider. In making his selection among members of the family of optimal redundancy allocations, the decision maker is assured that whatever the cost, weight, and volume used, maximum reliability is attained.

For clarity and ease of reference we list the exact models to be treated.

A. Parallel Redundancy, Given Set of Constraints. Stage i consists of $n_i + 1$ units in parallel, each of which has independent probability q_i $(0 < q_i < 1)$ of failing. Linear "cost" constraints exist on $\mathbf{n} = (n_1, n_2, \ldots, n_k)$ as follows:

$$\sum_{i=1}^{k} c_{ij} n_i \leq c_j, \qquad j = 1, 2, \ldots, r \tag{2.1}$$

where each $c_{ij} > 0$. Select a vector of nonnegative integers $\mathbf{n}$ to maximize system reliability $R(\mathbf{n})$, where in the present model $R(\mathbf{n})$ is given by

$$R(\mathbf{n}) = \prod_{i=1}^{k} (1 - q_i^{n_i+1}), \tag{2.2}$$

subject to (2.1).

To describe the next model we need the concept of an undominated redundancy allocation. We shall say $\mathbf{n}^0$ is *undominated* if $R(\mathbf{n}) > R(\mathbf{n}^0)$ implies $c_j(\mathbf{n}) > c_j(\mathbf{n}^0)$ for some j, whereas $R(\mathbf{n}) = R(\mathbf{n}^0)$ implies either $c_j(\mathbf{n}) > c_j(\mathbf{n}^0)$ for some j or $c_j(\mathbf{n}) = c_j(\mathbf{n}^0)$ for all j, where $c_j(\mathbf{n}) = \sum_{i=1}^{k} c_{ij} n_i$.

B. *Parallel Redundancy, No Specific Set of Constraints.* As in A, stage i consists of $n_i + 1$ units in parallel, each of which has independent probability q_i $(0 < q_i < 1)$ of failing. However, no specific constraints such as (2.1) are given. We wish to generate a family $\{\mathbf{n}^*\}$ of undominated allocations.

C. *Standby Redundancy, Given Set of Constraints.* A system is required to operate for the period $[0, t_0]$. When a component fails, it is immediately replaced by a spare component of the same type, if one is available. If no spare is available, system failure results. Only the spares originally provided may be used for replacements; that is, no resupply of spares can occur during $[0, t_0]$.

We now describe the original system before any replacements are made. The system consists of d_i "positions" or "sockets" each filled by a component of type i $(i = 1, 2, \ldots, k)$. The various components of a given type may be used at different levels of intensity and may be subject to different environmental stresses, so that for full generality we assume that the life of the jth component of type i (occupying position i, j, say) has probability distribution F_{ij}.

Each replacement has the same life distribution as its predecessor; component lives are mutually independent. Finally, position i, j is not required to be in constant operation throughout $[0, t_0]$ but rather is scheduled to operate for a period of duration $t_{ij} \leq t_0$.

What choice of n_i, the number of spares of type i initially provided $(i = 1, 2, \ldots, k)$, will maximize $R(\mathbf{n})$, the probability of continued system operation (that is, no shutdown resulting from shortage) during $[0, t_0]$, subject to the constraints (2.1)?

D. *Standby Redundancy, No Specific Set of Constraints.* The model is as in C, except that instead of desiring to maximize $R(\mathbf{n})$ subject to constraints such as (2.1), we wish to generate a family $\{\mathbf{n}^*\}$ of undominated allocations.

General solution of redundancy allocation problem

To discuss all four models conveniently, let us define $R(\mathbf{n})$ as the system reliability if n_i redundant units of type i are provided $(i = 1, 2, \ldots, k)$. In models A and B this means $n_i + 1$ units of component type i are actively operating in parallel; in models C and D this means n_i spares of component type i are in the spare parts kit initially. Since we are assuming a series-type structure for the basic system, we may write

$$R(\mathbf{n}) = \prod_{i=1}^{k} R_i(n_i), \tag{2.3}$$

where $R_i(n_i)$ is the reliability of the portion of the system using components of type i, assuming that n_i redundant units of type i are provided.

Let us study the relationship between models A and B (or between C and D). Under models B and D we are seeking a family of undominated redundancy allocations. Suppose in model B we find the *complete* family of undominated redundancy allocations, that is, the set consisting of all undominated redundancy allocations. Then given any particular problem of type A, we need only consider the complete family of undominated redundancy allocations; the solution must be a member of this family. In fact, the solution is the member of the complete family that achieves highest reliability among the members satisfying (2.1). Of course, a similar relationship exists between models C and D. Thus, solving model $B(D)$ essentially provides the solution to model $A(C)$.

We shall consider methods for generating complete families of undominated allocations and methods for generating incomplete families. Although the incomplete families are naturally less desirable, they can be generated much more rapidly and conveniently and provide approximate solutions when used under models A and C.

We start with a procedure for generating an incomplete family of undominated allocations which is intuitively quite reasonable in the single-constraint case. The underlying idea is that we shall construct successively larger redundancy allocations by adding one component at a time; the component we add will be the one that provides greatest improvement in system reliability per dollar spent (assuming the single constraint is on dollars spent). We shall show that each allocation so obtained is undominated if $\log R(\mathbf{n})$ is a concave function. We shall then show that for model A, $\log R(\mathbf{n})$ is concave (Section 3), and for model C, $\log R(\mathbf{n})$ is concave if each F_{ij} is IFR (Section 4).

From (2.3) we may write

$$\log R(\mathbf{n}) = \sum_{i=1}^{k} \log R_i(n_i). \tag{2.4}$$

Equation (2.4) is more convenient since each term of the sum depends only on a single variable. To determine the increment in the logarithm of system reliability resulting from adding a single unit of type i, we need only examine the increment in $\log R_i(n_i)$. Moreover, since $\log x$ is a monotone-increasing function of x, maximizing $\log R(\mathbf{n})$ is equivalent to maximizing $R(\mathbf{n})$.

Procedure 1 (single cost factor)

Assume $r = 1$. Start with the cheapest cost allocation $(0, 0, \ldots, 0)$. Obtain successively more expensive allocations as follows. If our present allocation is $\mathbf{n}$, we determine the index, say i_0, for which

$$\frac{1}{c_{i1}} [\log R_i(n_i + 1) - \log R_i(n_i)]$$

is maximum over $i = 1, 2, \ldots, k$. (If the maximum is achieved for more than one value of the index, choose the lowest among these.) Then the next allocation is $n_1, \ldots, n_{i_0-1}, n_{i_0} + 1, n_{i_0+1}, \ldots, n_k$; that is, we have added a single unit of the i_0th type to **n**.

Note that adding the most to log $R(\mathbf{n})$ per dollar spent is equivalent to multiplying $R(\mathbf{n})$ by the largest factor possible per dollar spent. We shall see in Theorem 2.1 of this chapter that procedure 1 generates only undominated allocations if log $R(\mathbf{n})$ is concave. But first let us consider the generalization of this procedure for the multiple cost case.

Procedure 1 (multiple cost factors)

Start with the allocation $(0, 0, \ldots, 0)$. Obtain successive allocations recursively as follows. If our present allocation is **n** we determine the index, say i_0, for which

$$\frac{1}{\sum_{j=1}^{r} a_j c_{ij}} [\log R_i(n_i + 1) - \log R_i(n_i)]$$

is maximum over $i = 1, 2, \ldots, k$. (If the maximum is achieved for more than one value of the index, we choose the lowest among these.) The $a_1, \ldots, a_r$ are nonnegative weights with $\sum_{j=1}^{r} a_j = 1$ whose choice will be discussed later. Then the next allocation is $n_1, \ldots, n_{i_0-1}, n_{i_0} + 1, n_{i_0+1}, \ldots, n_k$; that is, we have added a single unit of the i_0th type to **n**.

We choose in succession vectors $(a_1, \ldots, a_r)$, varying the a_j by some fixed increment, say Δ, until all choices from $(1, 0, 0, \ldots, 0)$ to $(0, 0, \ldots, 0, 1)$ have been exhausted.

Note that procedure 1 (single cost factor) is a special case of procedure 1 (multiple cost factors) obtained by letting $a_1 = 1, a_2 = 0, \ldots, a_r = 0$.

THEOREM 2.1. *If log $R(n)$ is concave, each redundancy allocation generated by procedure 1 is undominated.*

Proof. Let $\mathbf{n}^*$ be generated under procedure 1 by using the convex combination $a_1, \ldots, a_r$; let i_0 denote the index of the last component type added in arriving at $\mathbf{n}^*$ by procedure 1; and let

$$\lambda = \frac{1}{\sum_{j=1}^{r} a_j c_{i_0 j}} [\log R_{i_0}(n_{i_0}^*) - \log R_{i_0}(n_{i_0}^* - 1)]$$

Let **n** be any other allocation such that $R(\mathbf{n}) > R(\mathbf{n}^*)$. Designate the set of indices for which $n_i > n_i^*$ by I_1, and the set of indices for which

$n_i < n_i^*$ by I_2. Then we may write

$$0 < \log R(\mathbf{n}) - \log R(\mathbf{n}^*)$$

$$= \sum_{i \in I_1} [\log R_i(n_i) - \log R_i(n_i^*)]$$

$$- \sum_{i \in I_2} [\log R_i(n_i^*) - \log R_i(n_i)]$$

$$= \sum_{i \in I_1} \sum_{h=1}^{n_i - n_i^*} [\log R_i(n_i^* + h) - \log R_i(n_i^* + h - 1)]$$

$$- \sum_{i \in I_2} \sum_{h=0}^{n_i^* - n_i - 1} [\log R_i(n_i^* - h) - \log R_i(n_i^* - h - 1)]$$

$$\leq \sum_{i \in I_1} (n_i - n_i^*)[\log R_i(n_i^* + 1) - \log R_i(n_i^*)]$$

$$- \sum_{i \in I_2} (n_i^* - n_i)[\log R_i(n_i^*) - \log R_i(n_i^* - 1)]$$

by concavity of each $\log R_i(n)$. But the last expression does not exceed

$$\sum_{i \in I_1} (n_i - n_i^*) \lambda \sum_{j=1}^{r} a_j c_{ij} - \sum_{i \in I_2} (n_i^* - n_i) \lambda \sum_{j=1}^{r} a_j c_{ij}$$

since under procedure 1 successive increments in log reliability are decreasing. Since $\lambda > 0$,

$$0 < \sum_{j=1}^{r} a_j \sum_{i=1}^{k} c_{ij} n_i - \sum_{j=1}^{r} a_j \sum_{i=1}^{k} c_{ij} n_i^*.$$

It follows that for some index j,

$$\sum_{i=1}^{k} c_{ij} n_i > \sum_{i=1}^{k} c_{ij} n_i^*.$$

In a similar fashion, assuming $R(\mathbf{n}) = R(\mathbf{n}^*)$, we may prove that either $\sum_{i=1}^{k} c_{ij} n_i > \sum_{i=1}^{k} c_{ij} n_i^*$ for some j or $\sum_{i=1}^{k} c_{ij} n_i = \sum_{i=1}^{k} c_{ij} n_i^*$ for all j. Thus $\mathbf{n}^*$ is undominated. ∥

We emphasize that the family of undominated allocations obtained under procedure 1 when each $\log R_i(n)$ is concave is not necessarily complete, even if all convex combinations $a_1, \ldots, a_r$ are used. In practice, however, the undominated allocations obtained under procedure 1 are generally close enough to each other so that by selecting the appropriate member from this family a good approximation to the true solution in models A and C can be obtained.

Our next procedure also generates an incomplete family of undominated redundancy allocations when each $\log R_i(n)$ is concave. Heuristically, this procedure is based on the idea that an optimum balance has been struck in allocating among the different component types when the increments in log reliability per unit of "cost" (actually convex combination of costs) are the same for all component types, within the limitations of the discreteness of the decision variables $n_1, \ldots, n_k$.

Procedure 2

Let $\boldsymbol{\lambda} = (\lambda_1, \ldots, \lambda_r)$, where each $\lambda_j \geq 0$ (but not all $\lambda_j = 0$). For $i = 1, 2, \ldots, k$ obtain $n_i(\boldsymbol{\lambda})$ as the smallest integer m satisfying

$$\log R_i(m + 1) - \log R_i(m) < \sum_{j=1}^{r} \lambda_j c_{ij}. \tag{2.5}$$

THEOREM 2.2. *If $\log R(\mathbf{n})$ is concave, the $\mathbf{n}(\boldsymbol{\lambda})$ obtained in procedure 2 is undominated.*

The proof is similar to that of Theorem 2.1; for any $\mathbf{n}$ for which $R(\mathbf{n}) > R(\mathbf{n}(\boldsymbol{\lambda}))$, express $\log R(\mathbf{n}) - \log R(\mathbf{n}(\boldsymbol{\lambda}))$ as in Theorem 2.1 and infer that $c_j(\mathbf{n}) > c_j(\mathbf{n}(\boldsymbol{\lambda}))$ for some j, and similarly for $\mathbf{n}$ for which $R(\mathbf{n}) = R(\mathbf{n}(\boldsymbol{\lambda}))$. The assumption that $\log R(\mathbf{n})$ is concave is crucial.

Procedure 2 has the advantage that it is not necessary to generate each successively larger undominated redundancy allocation; rather, by selecting an appropriate $\boldsymbol{\lambda}$ we can arrive at a large redundancy allocation immediately. The trouble is that we do not know in advance which $\boldsymbol{\lambda}$ will yield a given reliability or cost vector. However, by selecting trial values of $\boldsymbol{\lambda}$ and then computing the resulting $\mathbf{n}(\boldsymbol{\lambda}), R(\mathbf{n}(\boldsymbol{\lambda})), c_1(\mathbf{n}(\boldsymbol{\lambda})), \ldots, c_r(\mathbf{n}(\boldsymbol{\lambda}))$, we can obtain redundancy allocations having reliability or cost vectors in a range of interest. In such trial calculations it is quite helpful to use the fact, evident from (2.5), that each $n_i(\boldsymbol{\lambda})$ is a decreasing function of each λ_j when $\log R(\mathbf{n})$ is concave. Consequently, $R(\mathbf{n}(\boldsymbol{\lambda}))$ and each $c_j(\mathbf{n}(\boldsymbol{\lambda}))$ are decreasing functions of $\lambda_1, \ldots, \lambda_r$.

By varying $\boldsymbol{\lambda}$ in procedure 2 we can generate all the undominated redundancy allocations obtained under procedure 1, except in the case of "ties." A tie occurs when under procedure 1 the maximum ratio is achieved simultaneously for several values of the index. The proof of this equivalence is similar to the proof of Theorem 2, Section 2.2 of Proschan (1960b). In many problems of type B or D, the most convenient way to generate undominated allocations of interest is to combine procedures 1 and 2 as follows. First, using procedure 2, arrive at an $\mathbf{n}(\boldsymbol{\lambda})$ whose reliability or cost vector is at the lower extreme of the range of interest. Then use procedure 1 to generate successively larger redundancy allocations by adding one unit at a time. Of course, in the case of several cost factors (that is, $r > 1$), it may be necessary to use several starting values under procedure 1 (corresponding to several convex combinations $a_1, \ldots, a_r$).

More specific details and examples of the various procedures in the separate cases of parallel and standby redundancy are given in the next two sections.

3. APPLICATION TO PARALLEL REDUNDANCY MODEL

Next we apply the general solution of the allocation problem obtained in Section 2 to models A and B, the parallel redundancy models. We shall present numerical examples to illustrate the details.

Recall that to establish the optimality of procedures 1 and 2 of Section 2, we must first establish that $\log R(\mathbf{n})$ is a concave function of $\mathbf{n}$, where $R(\mathbf{n})$ is given in (2.2). To this end, write

$$\Delta^2 \log R_i(n) = \Delta^2 \log (1 - q_i^{n+1}) = \log \frac{(1 - q_i^{n+3})(1 - q_i^{n+1})}{(1 - q_i^{n+2})^2},$$

where $\Delta f(n) = f(n+1) - f(n)$ in general. The denominator is larger than the numerator since

$$(1 - q_i^{n+2})^2 - (1 - q_i^{n+3})(1 - q_i^{n+1}) = q_i^{n+1}(q_i - 1)^2 > 0.$$

Hence $\Delta^2 \log R_i(n) < 0$, and so each $\log R_i$ is concave. Since the sum of concave functions is concave, it follows that

$$\log R(\mathbf{n}) = \sum_{i=1}^{k} \log R_i(n_i)$$

is concave.

Example 1. Procedure 1 (single cost factor). To illustrate the application of procedure 1 (single cost factor) to models A and B, we consider the example given in Kettelle (1962).

Consider a four-stage system with redundant units actively operating in parallel (models A and B); component cost and unreliability are given in Table 3.1.

TABLE 3.1

Stage, i	Cost, c_{i1}	Unreliability, q_i
1	1.2	.2
2	2.3	.3
3	3.4	.25
4	4.5	.15

REDUNDANCY OPTIMIZATION

We wish (a) to generate a family of undominated redundancy allocations (this corresponds to model B) and (b) to find the least-cost redundancy allocation for which system reliability is .99 or better (this corresponds to model A). (Note that the problem of achieving a given system reliability at minimum cost for redundancy is equivalent to the problem of maximizing system reliability while not exceeding a given cost for redundancy.)

(a) To generate a family of undominated redundancy allocations using procedure 1 (single cost factor), we start with the initial allocation (0, 0, 0, 0) which corresponds to no redundancy, that is, just the original single unit operating in each stage. To determine which component type to add as the first redundant unit, we compare

$$\frac{1}{c_{11}}[\log(1 - q_1^2) - \log(1 - q_1)] = .0660$$

$$\frac{1}{c_{21}}[\log(1 - q_2^2) - \log(1 - q_2)] = .0495$$

$$\frac{1}{c_{31}}[\log(1 - q_3^2) - \log(1 - q_3)] = .0285$$

$$\frac{1}{c_{41}}[\log(1 - q_4^2) - \log(1 - q_4)] = .0135.$$

Since the first ratio is the largest, we add a unit of the first type to arrive at our second redundancy allocation of (1, 0, 0, 0).

We show in detail the calculation of the next allocation. Now we compare

$$\frac{1}{1.2}[\log(1 - .2^3) - \log(1 - .2^2)] = .0119$$

$$\frac{1}{2.3}[\log(1 - .3^2) - \log(1 - .3)] = .0495$$

$$\frac{1}{3.4}[\log(1 - .25^2) - \log(1 - .25)] = .0285$$

$$\frac{1}{4.5}[\log(1 - .15^2) - \log(1 - .15)] = .0135.$$

Since the second ratio is the largest of the four, we add one unit of the second type to arrive at our next allocation (1, 1, 0, 0).

Proceeding in a similar fashion, adding one unit at a time, namely, the unit that increases log reliability most per unit of cost, we arrive at the family of undominated allocations displayed in Figure 3.1. We remind the reader that the family is not necessarily a complete one; in fact, using

Kettelle's algorithm (Procedure 3 of Section 5) in the same problem we obtain additional undominated allocations (see Table 5.3) interspersed among the ones shown in Figure 3.1. A comparison of Figure 3.1 with Table 5.3 confirms the fact that all the undominated allocations shown in

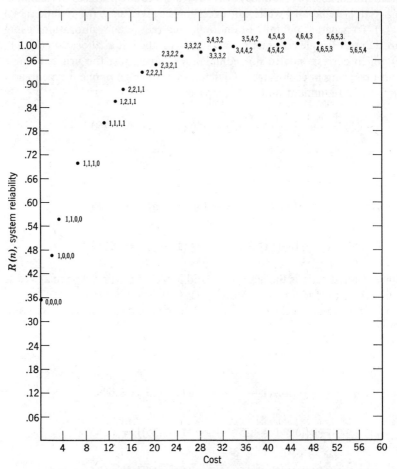

Fig. 3.1. Undominated allocations in single cost factor problem.

Figure 3.1 appear in Table 5.3 (keeping in mind that Table 5.3 is of more limited range).

(b) To obtain the member of the family shown in Figure 3.1 that achieves a reliability of .99 at minimum cost, we simply read off the curve the leftmost point with system reliability .99 or higher. This is the redundancy allocation (3, 4, 4, 2); that is, the system has in parallel 3, 4, 4, and 2 units of types 1, 2, 3, and 4 respectively. The corresponding cost is 35.4, and the system reliability actually attained is .992.

Note that the solution (3, 4, 4, 2) may be only approximate since the family of undominated allocations obtained in (a) is not complete. Since the cost of the allocation just to the left of (3, 4, 4, 2) in Figure 3.1 is 32.0, however, the error in our solution is less than 35.4 − 32.0 = 3.4. Actually, by using the Kettelle procedure of Section 5, the redundancy allocation achieving a reliability of .99 at minimum cost is (4, 4, 3, 2) at cost 33.2 (see Table 5.3).

Explicit formulas for procedure 2

In applying procedure 2 it is helpful to make use of the following explicit formula (3.1) for $n_i(\lambda)$ $(i = 1, \ldots, k)$. Recall from Equation (2.5) that by definition $n_i(\lambda)$ is the smallest integer n satisfying

$$\sum_{j=1}^{r} \lambda_j c_{ij} > \Delta \log R_i(n) = \log \frac{1 - q_i^{n+2}}{1 - q_i^{n+1}},$$

or exponentiating, satisfying

$$\exp\left\{\sum_{j=1}^{r} \lambda_j c_{ij}\right\} > \frac{1 - q_i^{n+2}}{1 - q_i^{n+1}} = 1 + \frac{p_i}{q_i^{-n-1} - 1},$$

or equivalently, satisfying

$$n > \frac{1}{\log q_i} \log \frac{\exp\left(\sum_{i=1}^{r} \lambda_j c_{ij}\right) - 1}{\exp\left(\sum_{j=1}^{r} \lambda_j c_{ij}\right) - q_i} - 1.$$

It follows that explicitly

$$n_i(\lambda) = \left[\frac{1}{\log q_i} \log \frac{\exp\left(\sum_{j=1}^{r} \lambda_j c_{ij}\right) - 1}{\exp\left(\sum_{j=1}^{r} \lambda_j c_{ij}\right) - q_i}\right], \quad i = 1, 2, \ldots, k, \quad (3.1)$$

where $[x]$ represents the largest integer not exceeding x.

Example 2. Procedure 1 (multiple cost factors). We consider the same four-stage system as in Example 1. However we assume in addition to component cost, that component weight is important. For completeness, we list the input data in Table 3.2.

Our problem is to generate a family of undominated allocations with cost not exceeding 56 and weight not exceeding 120.

Using procedure 1 we start with $a_1 = 0$, $a_2 = 1$ and generate in succession larger and larger undominated allocations by adding one unit at a time just as we did in Example 1. The last point so obtained is (5, 6, 5, 4).

TABLE 3.2

Stage, i	1	2	3	4
Component cost, c_{i1}	1.2	2.3	3.4	4.5
Component weight, c_{i2}	5	4	8	7
Unreliability, q_i	.2	.3	.25	.15

Using in turn weights $a_1 = .25$, $a_2 = .75$; $a_1 = .50$, $a_2 = .50$; $a_1 = .75$, $a_2 = .25$; and finally $a_1 = 1$, $a_2 = 0$, and generating successively larger redundancy allocations by adding one unit at a time following procedure 1, we obtain the set of redundancy allocations displayed in Figure 3.2.

Example 2. Procedure 2. Using procedure 2 find the redundancy allocation that achieves maximum system reliability at cost not exceeding 56 and weight not exceeding 120.

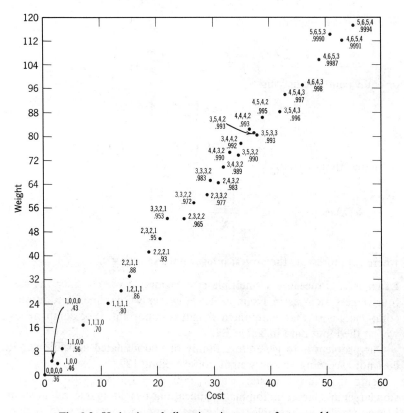

Fig. 3.2. Undominated allocations in two cost factor problem.

Using procedure 2 we select a grid of values λ_1, λ_2 and compute the corresponding $\mathbf{n}(\boldsymbol{\lambda})$ according to (3.1). Among the set of $\mathbf{n}(\boldsymbol{\lambda})$ obtained with cost not exceeding 56 and weight not exceeding 120, we find that the redundancy allocation (5, 6, 5, 4) achieves maximum reliability $R(\mathbf{n}) = .997$.

4. APPLICATION TO STANDBY REDUNDANCY MODEL

Finally, we apply the general solution of the allocation problem obtained in Section 2 to models C and D, the standby redundancy models. Again we present numerical examples to illustrate the details.

As before, in order to establish the optimality of procedures 1 and 2 we must first establish the conditions under which log $R(\mathbf{n})$ is concave, where $R(\mathbf{n})$ is the probability of continued system operation during $[0, t_0]$. It turns out that log $R(\mathbf{n})$ is *not* concave for *all* component failure distributions F_{ij} but, as shown in Theorem 4.1 of this chapter, is concave if each F_{ij} is IFR. (As usual, the very natural IFR assumption simplifies the solution considerably.)

First we must establish the relationship between $R(\mathbf{n})$ (call it system reliability) and the F_{ij} under models C and D. The lives of the component in position i, j and its successive replacements constitute a renewal process (as long as spares are available). Let N_{ij} equal the number of failures in position i, j during $[0, t_{ij}]$. Then $P[N_{ij} = n] = F_{ij}^{(n)}(t_{ij}) - F_{ij}^{(n+1)}(t_{ij})$. Moreover, $R_i(n)$, the probability that no shortage of spares of type i occurs during $[0, t_0]$, assuming n spares of type i are stocked, is given by

$$R_i(n) = P[N_{i1} + N_{i2} + \cdots + N_{id_i} \leq n]$$

or explicitly, by

$$R_i(n) = \sum_{n_1 + \cdots + n_{d_i} \leq n} \prod_{j=1}^{d_i} P[N_{ij} = n_j]. \quad (4.1)$$

Now we can show

THEOREM 4.1. If each F_{ij} is IFR, then log $R(\mathbf{n})$ is concave.

Proof. By Theorem 2.12 of Chapter 3, log $P[N_{ij} \leq n]$ is concave in n. By Theorem 5.3 of Chapter 2, this property is preserved under convolution so that $R_i(n) = P[N_{i1} + \cdots + N_{id_i} \leq n]$ is also log concave in n. It follows that

$$\log R(\mathbf{n}) = \sum_{i=1}^{k} \log R_i(n_i)$$

is concave in $\mathbf{n}$. ∥

Thus procedures 1 and 2 may be applied in the standby redundancy models C and D. The application is especially simple in the case of underlying exponential failure distributions as we shall see next.

Exponential failure distributions

Suppose $f_{ij}(t) = 1/(\mu_{ij})e^{-t/\mu_{ij}}$ for $t \geq 0$, 0 for $t < 0$. It follows that

$$P[N_{ij} = n] = e^{-t_{ij}/\mu_{ij}} \frac{(t_{ij}/\mu_{ij})^n}{n!},$$

a Poisson frequency function with parameter t_{ij}/μ_{ij} (Arrow, Karlin, and Scarf, 1958, p. 272).

Hence $$R_i(n) = e^{-\mu_i} \sum_{j=0}^{n} \frac{\mu_i^j}{j!} \qquad (4.2)$$

where $$\mu_i = \sum_{j=1}^{d_i} \frac{t_{ij}}{\mu_{ij}},$$

since the convolution of Poisson frequency functions is a Poisson frequency function with parameter given by the sum of the separate parameters (Cramér, 1946, p. 205).

Example 3 (Proschan, 1960b). A system consisting of an ultrahigh-frequency (UHF) receiving subsystem and a very high-frequency (VHF) receiving subsystem is to be placed in the field for a three-month period of experimentation. The tubes in the system have much higher probability of failure than do the remaining components. Thus we wish to determine spares kits for the tubes alone. We assume tube lives are stochastically independent and exponentially distributed, with mean lives given in Table 4.1.

TABLE 4.1

			Input data			Computed
i	Tube Type	Mean Life, hours	Cost per Tube, c_{i1}	Number in UHF, Scheduled for 332 hours of Use Each	Number in VHF, Scheduled for 2160 hours of Use Each	Expected Number of Failures, μ_i
1	Radechon	2500	$240	4	4	4.0
2	Memotron	4000	1025	2	5	2.9
3	Carcinotron	8000	1158	4	0	1.7
4	Traveling wave tube	6000	750	2	0	0.11

PROBLEM. Develop a family of undominated redundancy allocations (spares kits) having reliabilities starting at approximately .75 and ranging upward to a value close to 1.

Solution. First we compute the expected number of tubes of each type to be replaced during the period:

$$\mu_1 = \frac{1}{2500}(4 \times 332 + 4 \times 2160) = 4.0,$$

$$\mu_2 = \frac{1}{4000}(2 \times 332 + 5 \times 2160) = 2.9,$$

$$\mu_3 = \frac{1}{800}(4 \times 332) = 1.7,$$

$$\mu_4 = \frac{1}{6000}(2 \times 332) = .11.$$

These values appear in the last column of Table 4.1.

We shall use a combination of procedures 1 (single cost factor) and 2. To arrive at the first value of λ to use under procedure 2, we reason as follows. Since we are aiming for a system reliability of about .75, we would expect a reliability of roughly $.75^{1/4} = .93$ for a single tube type. From the Poisson tables (Molina, 1942) we find that corresponding to a Poisson parameter of $\mu_1 = 4.0$, $R_1(7) = .9489$, the nearest value to .93. Thus we may take as our starting value of λ a number slightly larger than

$$\frac{1}{240}[\log R_1(8) - \log R_1(7)] = .00005577.$$

With $\lambda = .000056$, using (2.5) we obtain

$$n_1(.000056) = 7$$
$$n_2(.000056) = 4$$
$$n_3(.000056) = 3$$
$$n_4(.000056) = 0.$$

To continue, we use procedure 1 (single cost factor), obtaining the points shown in Figure 4.1.

Example 4. We consider the same system as in Example 3, except that in addition to dollar cost we are interested in the total number of tubes carried as spares as a single rough measure of weight, volume, handling, storage, bookkeeping, etc., involved. This means that c_{i2} is taken as 1 for $i = 1, 2, \ldots, k$ and that $c_2(\mathbf{n}) = \sum_{i=1}^{k} n_i$. For completeness we list the input data in Table 4.2.

(*a*) Generate a family of undominated redundancy allocations (that is, spare part kits) starting with (0, 0, 0, 0) with cost not exceeding \$18,000 nor requiring more than 25 spare tubes.

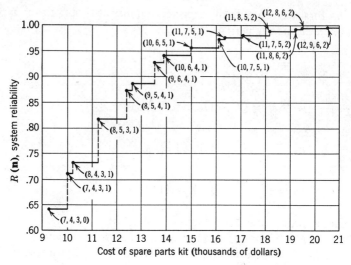

Fig. 4.1. Reliability-cost curve for undominated allocations.

Using procedure 1, we start with $a_1 = 0$, $a_2 = 1$. Using the fact that $R_i(n)$ is given by (4.4), we generate in succession the points $(0, 0, 0, 0)$, $(1, 0, 0, 0)$, $(1, 1, 0, 0)$, $(1, 1, 1, 0)$, etc., shown in Figure 4.2, with the reliability shown below the point. We stop when either the cost reaches $18,000 or the number of tubes reaches 25. Next we take $a_1 = .1$, $a_2 = .9$

TABLE 4.2

i	Tube Type	Cost per Tube, c_{i1}	c_{i2}	Expected Number of Failures, μ_i,
1	Radechon	$240	1	4.0
2	Memotron	1025	1	2.9
3	Carcinotron	1158	1	1.7
4	Traveling wave tube	750	1	0.11

and using procedure 1 generate in succession the points $(0, 0, 0, 0)$, $(1, 0, 0, 0)$, $(2, 0, 0, 0)$, $(3, 0, 0, 0)$, etc. We next compute by procedure 1 points corresponding to $a_1 = .2, .3, .4, .5, 1$, and of course $a_2 = 1 - a_1$. It turns out that no additional points are generated. The set of redundancy allocations shown in Figure 4.2 constitutes a family of undominated redundancy allocations.

(b) Find the redundancy allocation with cost not exceeding $18,000 and number of tubes not exceeding 25 which maximizes the probability of system survival.

REDUNDANCY OPTIMIZATION

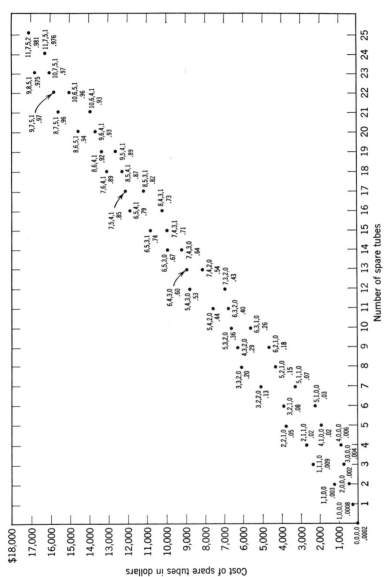

Fig. 4.2. Undominated allocations in two cost factor problem.

From Figure 4.2 we see that the redundancy allocation with cost not exceeding $18,000 and number of tubes not exceeding 25 which has maximum reliability calls for 11 radechons, 7 memotrons, 5 carcinotrons, and 1 traveling wave tube. The cost is $17,105, the number of tubes used is 25, and the reliability is .981. Note that the solution is only approximate since the family of undominated allocations may not be complete.

Gamma failure densities

If component life follows a gamma distribution, some of the graphs in Morrison and David (1960) may prove helpful in computing undominated redundancy allocations. This article contains graphs showing the probability of survival for various periods of time of a subsystem containing d like components when k spares are available, for various small values of d and k. One set of graphs covers the case of component failure density $f(t) = te^{-t}$, another the case of $f(t) = (t^3/3!)e^{-t}$.

5. COMPLETE FAMILIES OF UNDOMINATED ALLOCATIONS

Now let us consider a procedure for generating a *complete* family of undominated allocations in the case of a single cost factor, that is, $r = 1$. The method, from Kettelle (1962), was developed for parallel redundancy but clearly is applicable for standby redundancy. Basically, the Kettelle algorithm generates undominated redundancy allocations for successively larger subsystems from undominated allocations for small subsystems. The optimality of the Kettelle algorithm does *not* require that log $R(\mathbf{n})$ be concave. Specifically, we follow

Procedure 3 (Kettelle algorithm for single cost factor)

The simplest and clearest way to explain the Kettelle algorithm is to illustrate its operation in a numerical example. We take Example 1 of Section 3.

PROBLEM. Generate a *complete* family of undominated redundancy allocations, at least until system reliability reaches .99.

Solution. 1. Obtain a complete family of undominated redundancy allocations for the subsystem consisting of *stages 1 and 2 alone*. To obtain such a complete family set up Table 5.1. The column headings correspond to n_1, $n_1 c_{11}$, and $q_1^{n_1+1}$ whereas the row headings correspond to n_2, $n_2 c_{21}$, and $q_2^{n_2+1}$. The entries in the body of the table give the redundancy, cost, and unreliability for the subsystem consisting of stages 1 and 2. Thus the entry $(1, 0)$ corresponds to $n_1 = 1$, $n_2 = 0$, with cost 1.2, and subsystem unreliability $.04 + .30 - (.04)(.30) = .33$. The entry $(2, 1)$ corresponds to

$n_1 = 2$, $n_2 = 1$, with cost 4.7 and subsystem unreliability of $.09 + .008 = .098$. Throughout we approximate the unreliability of two subsystems combined by $Q_1 + Q_2$, where Q_1 is the unreliability of the first subsystem and Q_2 that of the second whenever the product term $Q_1 Q_2 < .01$. Kettelle (1962) shows that by dropping the product term $-Q_1 Q_2$ at every step, the error in Q, the unreliability attained for the entire system, is less than Q^2.

2. Start with (0, 0), the first undominated redundancy allocation.

3. The second undominated allocation is the *cheapest cost* entry with unreliability no higher than .44. Since (1, 0) has lower cost (1.2) than (0, 1) has (2.3), we select (1, 0).

4. In general, the *next* undominated allocation is the cheapest cost entry with unreliability no higher than the one just obtained. If several entries of identical cost qualify, choose the one with lowest unreliability. If several entries of identical cost and lowest unreliability qualify, choose the one with lowest component index. If the present entry is at the intersection of row i and column j, search for the next entry may be confined to the union of rows $1, 2, \ldots, i$ and columns $1, 2, \ldots, j$.

Sometimes the remainder of an entire row or column can be rejected as dominated. Let C_{ij} represent the cost of the entry of row i, column j, and Q_{ij} its unreliability; let C_{i0} represent the cost of the entry heading row i, and Q_{i0} its unreliability. If $Q_{ij} < Q_{i'0}$ where $i' < i$, all entries in row i' which cost more than C_{ij} are dominated. This follows since $Q_{i'k} > Q_{i'0} > Q_{ij}$. The same is true for columns. As an example, note that in Table 5.1 entry (0, 1) dominates every entry in the first row starting with (2, 0), since the unreliability .27 of (0, 1) is less than the unreliability .3 resulting from the use of 0 units in stage 2 and the cost 2.3 of (0, 1) is less than the cost of any entry in the first row starting with (2, 0).

In Table 5.1 the dominated partial rows and columns are indicated by shading.

5. For stages 3, 4, obtain in a similar fashion a family of undominated allocations (see Table 5.2), and similarly for stages 5, 6, etc.

6. Next, obtain the undominated allocations for the subsystem consisting of stages 1, 2, 3, 4 by using the undominated allocations of stages 1, 2 and the undominated allocations of stages 3, 4 as shown in Table 5.3. Follow the same principle as in step 4. The movements may be considerably more irregular as the subsystems become larger. In the illustrative four-stage problem, the resulting sequence of undominated allocations of Table 5.3 constitutes a complete family of undominated redundancy allocations.

7. In a similar fashion combine stages 5, 6, 7, 8, etc., until all stages have been combined. It is not necessary that subsystems being combined be of

TABLE 5.1
Stage 1

Redundancy Cost Unreliability	0 0 .2	1 1.2 .04	2 2.4 .008	3 3.6 .0016	4 4.8 .00032	5 6.0 .000064	6 7.2 .0000128	7 8.4 .00000256	8 9.6 .000000512
0 0 .3	0,0 0 .44	1,0 1.2 .33	2.4						
1 2.3 .09	0,1 2.3 .27	1,1 3.5 .13	2,1 4.7 .098	5.9					
2 4.6 .027	4.6	1,2 5.8 .067	2,2 7.0 .035	3,2 8.2 .0286	9.4				
3 6.9 .0081		8.1	2,3 9.3 .0161	3,3 10.5 .0097	4,3 11.7 .00842	12.9			
4 9.2 .00243			2,4 11.6 .01043	3,4 12.8 .00403	4,4 14.0 .00275	15.2			
5 11.5 .000729			13.9	3,5 15.1 .002329	4,5 16.3 .0010	5,5 17.5 .00079	18.7		
6 13.8 .0002187				17.4	4,6 18.6 .00054	5,6 19.8 .00028	6,6 21.0 .00023	22.2	
7 16.1 .00006561					20.9	5,7 22.1 .00012	6,7 23.3 .000078	7,7 24.5 .000068	25.7
8 18.4 .000019683						5,8 24.4 .000084	6,8 25.6 .000032	7,8 26.8 .000022	8,8 28.0 .000020
9 20.7 .0000059049						26.7	6,9 27.9 .000019	7,9 29.1	8,9 30.3

Stage 2

REDUNDANCY OPTIMIZATION

TABLE 5.2

		Stage							
Redundancy Cost Unreliability	0 0 .25	1 3.4 .0625	2 6.8 .0156	3 10.2 .003906	4 13.6 .00097	5 17.0 .00024	6 20.4 .000061	7 23.8 .000015	8 27.2 .000004
1 4.5 .0225	0,0 0 .36	1,0 3.4 .20	2,0 6.8 .17						
2 9.0 .003375	4.5	1,1 7.9 .08	2,1 11.3 .038	3,1 14.7 .026					
3 13.5 .00051		12.4	2,2 15.8 .019	3,2 19.2 .00728	4,2 22.6 .0044	5,2 26.0 .0036			
4 18.0 .000076			20.3	3,3 23.7 .0044	4,3 27.1 .00148	5,3 30.5 .00075	6,3 33.9 .00057	37.3	
5 22.5 .000011				28.2	31.6	5,4 35.0 .00032	6,4 38.4 .000137	7,4 41.8 .000091	45.2
6 27.0 .0000017						39.5	6,5 42.9 .000072	7,5 46.3 .000026	
							47.4		

Stage

equal size, although experience shows that wherever possible it is preferable to use subsystems of equal size. The resulting sequence represents a complete family of undominated redundancy allocations as proved in Theorem 5.3 of this chapter.

THEOREM 5.1. *The allocations obtained following procedure 3 constitute a complete family of undominated allocations.*

Proof. First we shall prove inductively that the allocations obtained following procedure 3 include all undominated allocations. For a single-stage system, the allocations obtained are *all* the allocations, and hence include all undominated allocations.

Assume then that the allocations obtained for a j stage system, where $j = 1, 2, \ldots, k - 1$, include all undominated allocations. Consider any allocation $\mathbf{n} = (n_1, \ldots, n_k)$. Then by inductive hypothesis $(n_1, \ldots, n_j)$, $j < k$, is dominated by or equals some undominated allocation $(n_1^*, \ldots, n_j^*)$ obtained by procedure 3; similarly $n_{j+1}, \ldots, n_k$ is dominated by or equals some undominated allocation $(n_{j+1}^*, \ldots, n_k^*)$ obtained by procedure 3. Thus either $n_i \equiv n_i^*$, $(i = 1, \ldots, j)$ or $R(n_1, \ldots, n_j) \leq R(n_1^*, \ldots, n_j^*)$, $c_1(n_1, \ldots, n_j) \geq c_1(n_1^*, \ldots, n_j^*)$ with one of the inequalities strict. Similarly either $n_i \equiv n_i^*$, $(i = j + 1, \ldots, k)$ or $R(n_{j+1}, \ldots, n_k) \leq R(n_{j+1}^*, \ldots, n_k^*)$, $c_1(n_{j+1}, \ldots, n_k) \geq c_1(n_{j+1}^*, \ldots, n_k^*)$ with one of the inequalities strict. It follows that either $n_i \equiv n_i^*$, $(i = 1, \ldots, k)$ or $R(\mathbf{n}) = R(n_1, \ldots, n_j)R(n_{j+1}, \ldots, n_k) \leq R(n_1^*, \ldots, n_j^*)R(n_{j+1}^*, \ldots, n_k^*) = R(\mathbf{n}^*)$ and that $c_1(\mathbf{n}) = c_1(n_1, \ldots, n_j) + c_1(n_{j+1}, \ldots, n_k) \geq c_1(n_1^*, \ldots, n_j^*) + c(n_{j+1}^*, \ldots, n_k^*) = c_1(\mathbf{n}^*)$ with one of the inequalities strict, so that $\mathbf{n}$ either equals or is dominated by $\mathbf{n}^*$.

It remains to be proved that $\mathbf{n}^*$ is dominated by or equals an allocation obtained under procedure 3. In the statements that follow we assume for ease of discussion that no ties in cost occur among different allocations. In case of ties a similar argument establishes the desired result. Assume $\mathbf{n}^*$ is *not* one of the allocations obtained under procedure 3. Let $\mathbf{n}'$ be the first allocation generated under procedure 3 from combining stages $1, \ldots, j$ and stages $j + 1, \ldots, k$ whose cost exceeds $c_1(\mathbf{n}^*)$, and let $\mathbf{n}^0$ be the last allocation whose cost is less than $c_1(\mathbf{n}^*)$. Then $R(\mathbf{n}^0) > R(\mathbf{n}^*)$ since if $R(\mathbf{n}^0) \leq R(\mathbf{n}^*)$, then $\mathbf{n}^*$ would lie between $\mathbf{n}^0$ and $\mathbf{n}'$ under procedure 3, contrary to assumption. Thus $\mathbf{n}^*$ is dominated by $\mathbf{n}^0$ and the induction is complete.

We must also prove that every allocation obtained by using procedure 3 is undominated. Suppose $\mathbf{n}^0$ is an allocation obtained by using procedure 3. Then if $\mathbf{n}_0$ is dominated by any allocation, it must also be dominated by an undominated allocation. But we have just proved that all undominated allocations are obtained by procedure 3. Thus $\mathbf{n}^0$ is dominated by, say,

TABLE 5.3

Undominated allocations for stages 1, 2

Redundancy Cost Unreliability	0,0 0 .44	1,0 1.2 .33	0,1 2.3 .27	1,1 3.5 .13	2,1 4.7 .098	1,2 5.8 .067	2,2 7.0 .035	3,2 8.2 .0286	2,3 9.3 .0161	3,3 10.5 .0097	4,3 11.7 .00842	3,4 12.8 .00403	4,4 14.0 .00275	3,5 15.1 .00233	4,5 16.3 .0010	5,5 17.5 .00079	4,6 18.6 .00054	5,6 19.8 .00028
0,0 0 .36	0,0,0,0 0 .64	1,0,0,0 1.2 .57	0,1,0,0 2.3 .54	1,1,0,0 3.5 .45	2,1,0,0 4.7 .42	1,2,0,0 5.8 .41	7.0											
1,0 3.4 .20	0,0,1,0 3.4 .55	1,0,1,0 4.6 .46	0,1,1,0 5.7 .42	1,1,1,0 6.9 .30	2,1,1,0 8.1 .28	1,2,1,0 9.2 .26	2,2,1,0 10.4 .23	11.6										
2,0 6.8 .17	6.8	8.0	0,1,2,0 9.1 .40	1,1,2,0 10.3 .28	2,1,2,0 11.5 .25	1,2,2,0 12.6 .23	13.8											
1,1 7.9 .08			10.2	1,1,1,1 11.4 .20	2,1,1,1 12.6 .178	1,2,1,1 13.7 .147	2,2,1,1 14.9 .115	3,2,1,1 16.1 .109	2,3,1,1 17.2 .096	18.4								
2,1 11.3 .038				1,1,2,1 14.8 .168	2,1,2,1 16.0 .136	1,2,2,1 17.1 .105	2,2,2,1 18.3 .073	3,2,2,1 19.5 .067	2,3,2,1 20.6 .054	3,3,2,1 21.8 .048	4,3,2,1 23.0 .046	3,4,2,1 24.1 .042	25.3					
3,1 14.7 .026				18.2	19.4	20.5	2,2,3,1 21.7 .061	3,2,3,1 22.9 .055	2,3,3,1 24.0 .042	3,3,3,1 25.2 .036	4,3,3,1 26.4 .034	3,4,3,1 27.5 .030	28.7					
2,2 15.8 .019							2,2,2,2 22.8 .054	3,2,2,2 24.0 .048	2,3,2,2 25.1 .035	3,3,2,2 26.3 .029	4,3,2,2 27.5 .027	3,4,2,2 28.6 .023	29.8					
3,2 19.2 .00728							26.2	3,2,3,2 27.4 .036	2,3,3,2 28.5 .0234	3,3,3,2 29.7 .0170	4,3,3,2 30.9 .0157	3,4,3,2 32.0 .0113	4,4,3,2 33.2 .0100	3,5,3,2 34.3 .0096	4,5,3,2 35.5 .0083	36.7		
4,2 22.6 .0044								30.8	2,3,4,2 31.9 .0205	3,3,4,2 33.1 .0141	4,3,4,2 34.3 .0128	3,4,4,2 35.4 .0084	4,4,4,2 36.6 .0071	3,5,4,2 37.7 .0067	4,5,4,2 38.9 .0054	5,5,4,2 40.1 .0052	41.2	
5,2 26.0 .0036										35.3	36.5	37.7	3,4,5,2 38.8 .0076	4,4,5,2 40.0 .0064	3,5,5,2 41.1 .0059	4,5,5,2 42.3 .0046	43.5	
4,3 27.1 .00148											3,4,4,3 39.9 .0055	4,4,4,3 41.1 .0043	3,5,4,3 42.2 .0038	4,5,4,3 43.4 .0025	5,5,4,3 44.6 .0023	4,6,4,3 45.7 .0020	5,6,4,3 46.9 .00176	
5,3 30.5 .00075												43.5	44.5	45.6	4,5,5,3 46.8 .0018	5,5,5,3 48.0 .00154	4,6,5,3 49.1 .00129	50.3 .00103
6,3 33.9 .00057															51.4 .00136			

Undominated allocations for stages 3, 4

$\mathbf{n}'$, also obtained under procedure 3. This is a contradiction since if $\mathbf{n}^0$ and $\mathbf{n}'$ are both obtained under procedure 3, either (a) $R(\mathbf{n}^0) < R(\mathbf{n}')$ and $c_1(\mathbf{n}^0) < c_1(\mathbf{n}')$ or (b) $R(\mathbf{n}^0) = R(\mathbf{n}')$ and $c_1(\mathbf{n}^0) = c_1(\mathbf{n}')$ or (c) $R(\mathbf{n}^0) > R(\mathbf{n}')$ and $c_1(\mathbf{n}^0) > c_1(\mathbf{n}')$. ‖

Example 1. Procedure 3. As before, we wish to find the least-cost redundancy allocation that achieves system reliability of .99 or better for the four-stage system of Example 1 of Section 3.

From Table 5.3 we simply read off the first undominated allocation with system unreliability not exceeding .01; this is the allocation (4, 4, 3, 2) with unreliability .0100 and cost 33.2. Actually, it is not necessary to generate the entire sequence of undominated allocations starting with (0, 0, 0, 0). Instead we could construct abbreviated versions of Tables 5.1, 5.2, and 5.3 so that unreliabilities greater than .01 would not appear. Thus the first column heading in Table 5.1 would correspond to $n_1 = 2$ with stage 1 unreliability .008, and the first row heading would correspond to $n_2 = 3$ with stage 2 unreliability .0081. Similarly, the first column heading in Table 5.2 would correspond to $n_3 = 3$ with stage 3 unreliability .0039, and the first row heading would correspond to $n_4 = 2$ with stage 4 unreliability .003375. Fnally, the first column heading in Table 5.3 would correspond to $n_1 = 3$, $n_2 = 3$ with unreliability .0097, and the first row heading would correspond to $n_3 = 3$, $n_2 = 2$ with unreliability .00728.

At present we are developing a generalization of Kettelle's procedure which generates a complete family of undominated redundancy allocations when any number of cost factors are present ($r > 1$). As in the Kettelle procedure, we generate undominated allocations for subsystems; from these we generate undominated allocations for larger subsystems, continuing until in the final combination of subsystems, undominated allocations for the entire system are generated. A detailed account is given in Proschan and Bray (1963).

6. OPTIMAL REDUNDANCY ASSUMING TWO TYPES OF FAILURE

In this section we treat the problem of achieving optimum redundancy for certain types of systems assuming that failure may take either one of two forms. We assume that all the components have the same reliability and that no constraints are imposed on the number of components to be used. Thus the model differs quite significantly from the allocation models studied in Sections 2, 3, and 4. For some historical background on this model refer back to Section 1. The material of this section is based on Barlow, Hunter, and Proschan (1963).

For example, consider a network of relays arranged so that there are m subsystems in parallel, each subsystem consisting of n relays in series; we shall call such an arrangement a parallel-series arrangement. An open-circuit failure of a single relay in each subsystem would make the system unresponsive; a short-circuit failure of all the relays in any subsystem would also make the entire system unresponsive. Analogously, we may consider a system consisting of n subsystems in series, each subsystem containing m relays in parallel (a series-parallel arrangement). A short-circuit failure of a single relay in each subsystem makes the system unresponsive; open-circuit failure of all the relays in a subsystem also makes the system unresponsive. As a third example (oversimplified), one of several airplane autopilots in parallel may fail by no longer functioning or by giving a command to make an extreme, destructive maneuver: the first type of failure simply leaves one less autopilot available for guidance; the second type causes system failure. Note that in all these examples adding components will make system failure of one type more likely at the same time that it reduces the chance of system failure of the other type. Thus a valid problem exists in determining the optimum number of components to use.

We shall treat the general model that may be abstracted from these separate examples; however, to keep the discussion simple and readily interpretable, we shall discuss the general model as it is exemplified by the parallel-series arrangement of relays. We emphasize, however, that the solution applies also to the series-parallel arrangement, to systems other than electrical circuits, etc. The only common requirement is that the system fail if (*a*) a *single* component in each subsystem experiences a failure of one type (for example, in the parallel-series arrangement of relays an open-circuit failure) or if (*b*) *each* component in a single subsystem experiences a failure of the other type (a short-circuit failure in the parallel-series arrangement of relays).

We shall treat two versions of the problem. First we shall show how to maximize system reliability. Second we shall show how to maximize expected system life.

Maximizing system reliability

We assume a parallel-series arrangement, m subsystems in parallel, each containing n relays in series. Each relay has independent probability $p_1 > 0$ of experiencing an open-circuit failure, $p_2 > 0$ of experiencing a closed-circuit failure, where $p_1 + p_2 < 1$. A relay can fail in only one way and, having failed, cannot recover and is not replaced. For given n we wish to find the value of m maximizing system reliability, that is, the probability that the system is functioning.

First note that system reliability R is given by

$$R = (1 - p_2^n)^m - (1 - q_1^n)^m, \tag{6.1}$$

where $q_i = 1 - p_i$, $i = 1, 2$. Here $1 - p_2^n$ represents the probability that not all the components in a given subsystem have experienced short-circuit failure; $(1 - p_2^n)^m$ represents the probability that none of the subsystems has experienced short-circuit failure; $1 - q_1^n$ is the probability that at least one component in a given subsystem has experienced open-circuit failures; and $(1 - q_1^n)^m$ is the probability that every subsystem has at

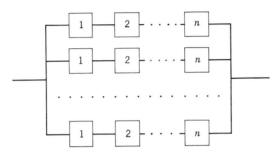

Fig. 6.1.

least one component which has experienced open-circuit failure. Thus system reliability is represented as the probability that the system does not experience short-circuit failure diminished by the probability that it does experience open-circuit failure, using the fact that system failures of the two types are mutually exclusive.

It is worth noting that we cannot seek a pair m, n maximizing system reliability R, since R can be made arbitrarily close to 1 by appropriate choice of m, n. To see this, let

$$a = \frac{\log p_2 - \log q_1}{\log p_2 + \log q_1}, \quad M_n = p_2^{-n/(1+a)}, \quad m_n = [M_n],$$

where $[x]$ equals the largest integer not exceeding x. For given n, take $m = m_n$; the corresponding system reliability R is given by

$$R = (1 - p_2^n)^{m_n} - (1 - q_1^n)^{m_n}.$$

Thus

$$\lim_{n \to \infty} R = \lim_{n \to \infty} [(1 - M_n^{-(1+a)})^{m_n} - (1 - M_n^{-(1-a)})^{m_n}]$$

$$= \lim_{M_n \to \infty} [(1 - M_n^{-(1+a)})^{M_n} - (1 - M_n^{-(1-a)})^{M_n}] = 1$$

since $a > 0$.

For fixed n, however, we can find the value of m maximizing system reliability $R(m)$, as shown in

THEOREM 6.1. For fixed n, p_1, p_2, the maximum value of $R(m)$ is attained at $m^* = [m_0] + 1$, where

$$m_0 = n \frac{\log q_1 - \log p_2}{\log(1 - p_2^n) - \log(1 - q_1^n)}. \tag{6.2}$$

If m_0 is an integer, m_0 and $m_0 + 1$ both maximize $R(m)$.

Proof. Using (6.1), we set

$$0 = \Delta R(m) = R(m+1) - R(m) = -p_2^n(1 - p_2^n)^m + q_1^n(1 - q_1^n)^m.$$

If we define m_0 as the solution, then m_0 is given by (5.2).

Note that for $m < m_0$, $\Delta R(m) > 0$, whereas for $m > m_0$, $\Delta R(m) < 0$; this is a consequence of the fact that $1 - p_2^n > 1 - q_1^n$ (since $q_1 > p_2$). Thus $[m_0] + 1$ maximizes $R(m)$. Finally, if m_0 is an integer, then $\Delta R(m_0) = 0$, so that $R(m_0) = R(m_0 + 1)$, and both m_0 and $m_0 + 1$ maximize $R(m)$. ∥

It is of value to understand how the solution m^* depends on the various parameters. First we study the behavior of m_0 as we vary n. We prove

THEOREM 6.2. m_0/n is a strictly increasing function of n.

Proof. From (6.2) we see that it is sufficient to prove that

$$(1 - p_2^n)/(1 - q_1^n)$$

is a strictly decreasing function of n or, equivalently, that

$$a_{n-1} = \frac{1 + p_2 + \cdots + p_2^{n-1}}{1 + q_1 + \cdots + q_1^{n-1}}$$

is a strictly decreasing function of n. We need only show

$$\operatorname{sgn}\{a_n - a_{n+1}\} = \operatorname{sgn} \begin{vmatrix} 1 + p_2 + \cdots + p_2^n & p_2^{n+1} \\ 1 + q_1 + \cdots + q_1^n & q_1^{n+1} \end{vmatrix}$$

$$= \operatorname{sgn}\left\{q_1^{n+1} \sum_{i=0}^{n} p_2^i - p_2^{n+1} \sum_{i=0}^{n} q_1^i\right\} > 0$$

which is obvious upon a term-by-term comparison since $p_2 < q_1$. ∥

From (6.2) we see that as $n \to \infty$

$$\frac{m_0}{n} \approx \left(\log \frac{q_1}{p_2}\right)\left(\frac{1}{q_1}\right)^n,$$

which of course approaches ∞ exponentially fast, since $1 > q_1 > p_2$.

We next prove

REDUNDANCY OPTIMIZATION

THEOREM 6.3. $m_0 < 1, = 1,$ or > 1 according as $q_1{}^n + p_2{}^n > 1, = 1,$ or < 1.

Proof. From (6.2) we see that $m_0 < 1, = 1,$ or > 1 according as

$$\frac{q_1{}^n}{p_2{}^n} - \frac{1 - p_2{}^n}{1 - q_1{}^n} < 0, = 0, \text{ or } > 0,$$

that is, according as $q_1{}^n + p_2{}^n > 1, = 1,$ or < 1. ‖

Making the following additional assumptions we obtain further qualitative results on the behavior of m_0:

Assumptions A. Each component has independent probability $F(t)$ of failing by time t, $(0 \leq t < \infty)$ with $F(0) = 0$. Given that failure has occurred, the conditional probability of open-circuit failure is p, of closed-circuit failure is q, $(p + q = 1)$, independently of the time of failure.

THEOREM 6.4. m_0 is a strictly increasing function of p for $0 \leq p \leq 1$, $0 < F(t) < 1$.

Proof. Since $p_1 = pF(t), p_2 = qF(t)$, (6.2) becomes

$$m_0 = n \frac{\log \{1 - pF(t)\} - \log \{qF(t)\}}{\log \{1 - [qF(t)]^n\} - \log \{1 - [1 - pF(t)]^n\}}. \quad (6.3)$$

The numerator is strictly increasing in p since $(1 - pF)/(F - pF)$ is. To verify that the denominator is strictly decreasing in p, we write

$$\frac{1 - (qF)^n}{1 - (1 - pF)^n} = \frac{1 - (qF)^n}{1 - (\bar{F} + qF)^n}$$

$$= \frac{1}{1 - \dfrac{\bar{F}^n + n\bar{F}^{n-1}(qF) + \cdots + n\bar{F}(qF)^{n-1}}{1 - (qF)^n}},$$

where $\bar{F} = 1 - F$. Since

$$\frac{\bar{F}^n + n\bar{F}^{n-1}(qF) + \cdots + n\bar{F}(qF)^{n-1}}{1 - (qF)^n}$$

is strictly decreasing in p, we conclude that the denominator of (6.3) is strictly decreasing in p. The result then follows. ‖

THEOREM 6.5. $m_0 \to n$ as $t \to 0$; $m_0 \to (1 - q^n)/q^n$ as $t \to \infty$.

Proof. The results follow immediately using L'Hospital's rule on (6.3) considering m_0 as a function of F and then using the continuity of F at the origin and at infinity. ‖

THEOREM 6.6. For $n = 1, \frac{1}{2} < p < 1, m_0$ is a strictly increasing function of F.

Proof. Let $y = \bar{F}/F$. From (6.3),

$$m_0 = \frac{\log(1 + y/q)}{\log(1 + y/p)}.$$

Here dm_0/dy has the same sign as does $(p + y)\log(1 + y/p) - (q + y)\log(1 + y/q) = W(y)$. But $W(0) = 0$ and $dW/dy = \log(1 + y/p) - \log(1 + y/q) < 0$ for $y > 0$ since $p > q$. Thus $W(y) < 0$ for $y > 0$, and so m_0 is strictly decreasing in y. Since y is strictly decreasing in F, we conclude that m_0 is strictly increasing in F. ∥

Maximizing expected system life

We have seen how to determine the number of subsystems required to maximize the probability of successful system operation until any specified time, given the relay life distribution $F(t)$. Often, however, the desired operating time is indeterminate. In such cases it is reasonable to maximize the expected time until system failure.

We next derive general expressions for the expected time until system failure for the case $n = 1$ and show how to choose the number of components required to maximize it. Using these results we obtain explicit answers for the case of exponential failure distributions and uniform failure distributions.

As before, we assume m subsystems in parallel; but we now specialize to the case in which each subsystem consists of one relay. Let the previously given Assumptions A hold; that is, each component life is independently distributed according to $F(t)$ (nondegenerate), with the conditional probability of an open-circuit failure given by $p > 0$ and of a closed-circuit failure by $q = 1 - p$. Letting $L_m(p)$ be the expected life of the system, we have

$$L_m(p) = \sum_{i=1}^{m-1} \mu_{i,m} p^{i-1} q + \mu_{m,m} p^{m-1} \tag{6.4}$$

where $\mu_{i,m}$ equals the expected value of the ith smallest observation in a sample of m observations independently selected from $F(t)$. Equation (6.4) simply states that system failure can occur by having $i - 1$ open-circuit failures preceding the first short-circuit failure, the conditional expected system life in this case being $\mu_{i,m}$ ($i = 1, 2, \ldots, m - 1$); in addition, we may have $m - 1$ open-circuit failures preceding the final relay failure which may be of either type, the conditional expected system life being $\mu_{m,m}$.

Now we shall show how to determine the number m of relays needed to maximize $L_m(p)$.

Form the increment

$$\Delta L_m(p) = L_{m+1}(p) - L_m(p). \tag{6.5}$$

REDUNDANCY OPTIMIZATION

In what follows we shall find for given p the value m such that for smaller values $\Delta L_m(p) > 0$ whereas for remaining values $\Delta L_m(p) < 0$. Thus m will maximize expected system life for the given value of p. First we prove

LEMMA 6.7. $\Delta L_m(p) = 0$ has one root r_m in $[0, 1]$. For $0 \leq p < r_m$, $\Delta L_m(p) < 0$, whereas for $1 \geq p > r_m$, $\Delta L_m(p) > 0$.

Proof. From (6.4) and Assumptions A,

$$L_m(p) = \int_0^\infty \{[1 - qF(t)]^m - [pF(t)]^m\}\, dt,$$

so that

$$\Delta L_m(p) = -\int_0^\infty [1 - qF(t)]^m qF(t)\, dt + \int_0^\infty [pF(t)]^m [1 - pF(t)]\, dt. \quad (5.6)$$

Note that

$$\Delta L_m(0) = -\int_0^\infty [1 - F(t)]^m F(t)\, dt < 0$$

and

$$\Delta L_m(1) = \int_0^\infty [F(t)]^m [1 - F(t)]\, dt > 0$$

when $m > 0$. Hence $\Delta L_m(p)$ has at least one root in $0 < p < 1$.

Moreover, from (6.4) we may write

$$\Delta L_m(p) = \sum_{i=1}^m a_i p^{i-1} q + a_{m+1} p^m$$

where
$$a_i = \begin{matrix} \mu_{i,m+1} - \mu_{i,m}, & \text{for } i = 1, 2, \ldots, m \\ \mu_{m+1,m+1} - \mu_{m,m}, & \text{for } i = m + 1. \end{matrix}$$

Let
$$f(p) = \sum_{i=1}^m \frac{a_i q}{p^{m-i+1}}$$

and note that $\Delta L_m(p) = p^m[f(p) + a_{m+1}]$ has as many roots in $0 < p < 1$ as $f(p) + a_{m+1}$. Since $a_i \leq 0$ ($i = 1, 2, \ldots, m$) and $(1 - p)/p^{m-i}$ is decreasing, $f(p) + a_{m+1}$ has at most one root in $0 < p < 1$. Hence $\Delta L_m(p)$ has at most one root in $0 < p < 1$. ∥

We shall make use of the monotonic character of the roots $r_1, r_2, \ldots$ as stated in

LEMMA 6.8. $.5 = r_1 < r_2 < r_3 < \cdots$

Proof. From (6.6) we have

$$\Delta L_m(p) = -\int_0^\infty [1 - qF(t)]^m qF(t)\, dt + \int_0^\infty [pF(t)]^m [1 - pF(t)]\, dt$$

Hence

$$\Delta L_{m+1}(r_m) = -\int_0^\infty [1 - (1 - r_m)F(t)]^m(1 - r_m)F(t)[1 - (1 - r_m)F(t)]\,dt$$

$$+ \int_0^\infty [r_m F(t)]^m[1 - r_m F(t)][r_m F(t)]\,dt$$

$$< r_m\left\{-\int_0^\infty [1 - (1 - r_m)F(t)]^m(1 - r_m)F(t)\,dt\right.$$

$$\left. + \int_0^\infty [r_m F(t)]^m[1 - r_m F(t)]\right\}dt = r_m \Delta L_m(r_m) = 0.$$

By Lemma 6.7 $r_{m+1} > r_m$ for $m = 1, 2, \ldots$.

Finally, note that

$$\Delta L_1(p) = -q\int_0^\infty [1 - qF(t)]F(t)\,dt$$

$$+ p\int_0^\infty F(t)[1 - F(t)]\,dt = 0$$

has the root $r_1 = .5$. ‖

Using these lemmas we finally obtain the value of m maximizing expected system life as a function of p.

THEOREM 6.9. For $r_{m-1} < p < r_m$, the number of components yielding maximum expected system life is m. For $p = r_m$, both m and $m + 1$ yield the maximum.

Proof. For $p < r_m$, $\Delta L_i(p) < 0$ for $i = m, m + 1, m + 2, \ldots$, implying m is preferable to $m + 1, m + 2, \ldots$; for $p > r_{m-1}$, $\Delta L_i(p) > 0$ for $i = 1, 2, \ldots, m - 1$, implying m is preferable to $1, 2, \ldots, m - 1$. Thus for $r_{m-1} < p < r_m$, m maximizes expected system life.

For $p = r_m$, $\Delta L_m(p) = 0$, implying both m and $m + 1$ yield the same value of expected system life. By the same arguments as before, m and $m + 1$ are preferable to all other values. ‖

In particular, note that since $r_1 = .5$, then for $p \leq .5$, the optimum number of components is 1.

Exponential failure distribution

Theorem 6.9 gives us a procedure for determining the number of parallel components maximizing expected system life. We illustrate for the case of $F(t) = 1 - e^{-\lambda t}$ $(\lambda > 0, 0 \leq t < \infty)$.

In this case we may write down directly

$$L_m(p) = \frac{1}{\lambda}\sum_{i=0}^{m-1} \frac{p^i}{m - i}$$

since $1/m\lambda$ is the expected time until the first failure, $p/(m-1)\lambda$ is the expected additional time elapsed until the second failure multiplied by the probability that the first failure is open circuit, $p^2/(m-2)\lambda$ is the expected additional time elapsed from the second until the third failure multiplied by the probability that the first two failures are open circuit, etc. Hence

$$\Delta L_m(p) = \frac{1}{\lambda}\left(\sum_{i=0}^{m} \frac{p^i}{m+1-i} - \sum_{i=0}^{m-1} \frac{p^i}{m-i}\right) = 0$$

TABLE 6.1*

COMPONENTS SUBJECT TO EXPONENTIAL FAILURE

Range of p	Number of Components Required to Maximize Expected System Life
$0 \leq p \leq .5$	1
$.500 < p \leq .728$	2
$.728 < p \leq .825$	3
$.825 < p \leq .875$	4
$.875 < p \leq .904$	5
$.904 < p \leq .923$	6
$.923 < p \leq .937$	7
$.937 < p \leq .946$	8
$.946 < p \leq .954$	9
$.954 < p \leq .960$	10
$.960 < p \leq 1$	>10

* From Barlow, Hunter, and Proschan, Optimum redundancy when components are subject to two kinds of failure, *J. Soc. Indust. Appl. Math.*, March 1963, Vol. 11, No. 1, pp. 64–73.

will yield the roots r_m needed to determine the maximizing values of m. Note that the roots are independent of λ. In fact, from (6.4) and (6.5) we see that the r_m and hence the maximizing m are independent of any scale factor in the component failure distribution.

Table 6.1 summarizes the numerical results.

Uniform failure distribution

Since $\mu_{1,m} = a/(m+1)$ and $\mu_{i,m} - \mu_{i-1,m} = a/(m+1)$ for $i = 2, 3, \ldots, m$ for the uniform distribution on $[0, a]$ (Cramér, 1946, p. 372), we may write

$$L_m(p) = a \sum_{i=1}^{m} \frac{p^{i-1}}{m+1}.$$

Note that $a/(m+1)$ is the expected time until the first failure, $[a/(m+1)]p$

is the expected additional time until the second failure multiplied by the probability that the first failure is open circuit, $[a/(m + 1)]p^2$ is the expected additional time between the second and third failure multiplied by the probability that the first two failures are open circuit, etc.

Hence the roots of

$$\Delta L_m(p) = \frac{a}{q}\left(\frac{1 - p^{m+1}}{m + 1} - \frac{1 - p^m}{m + 1}\right) = 0$$

are independent of a as noted more generally earlier. Rewriting, we have

$$(m + 1)p^{m+1} - (m + 2)p^m + 1 = 0. \tag{6.7}$$

Solving (6.7) yields Table 6.2.

TABLE 6.2*

COMPONENTS SUBJECT TO UNIFORM FAILURE DISTRIBUTION

Range of p	Number of Components Required to Maximize Expected System Life
$0 \leq p \leq .5$	1
$.5 < p \leq .77$	2
$.77 < p \leq .87$	3
$.87 < p \leq .92$	4
$.92 < p \leq .94$	5
$.94 < p \leq .96$	6
$.96 < p \leq .968$	7
$.968 < p \leq .974$	8
$.974 < p \leq .979$	9
$.979 < p \leq .984$	10
$.984 < p \leq 1$	>10

* From Barlow, Hunter, and Proschan, Optimum redundancy when components are subject to two kinds of failure, *J. Soc. Indust. Appl. Math.*, March 1963, Vol. 11, No. 1, pp. 64–73.

Similar tabulations can be obtained for other component distributions. For example, for the Weibull distribution

$$F(t) = 1 - e^{-t^\alpha}, \quad t > 0, \alpha > 0$$

Lieblein (1953) gives

$$\mu_{i,m} = \frac{m!}{(i - 1)!(m - i)!} \Gamma\left(1 + \frac{1}{\alpha}\right) \sum_{\mu=0}^{i-1} (-1)^\mu \binom{i - 1}{\mu}(m + \mu - i + 1)^{-1-1/\alpha}.$$

As before, we may solve for the r_m using this expression for $\mu_{i,m}$, and thereby obtain the optimizing values of m.

Clark and Williams (1958) give the following expansion for $\mu_{i,m}$ for a general distribution function $F(t)$ with density $f(t)$ assuming only that the required derivatives exist:

$$\mu_{i,m} = \frac{1}{F\left(\dfrac{i}{m+1}\right)} - \frac{f'}{2f^3} \cdot \frac{i(m+1-i)}{(m+1)^2(m+2)}$$
$$+ \frac{3f'^2 - ff''}{6f^5} \cdot \frac{2i(m+1-i)(m+1-2i)}{(m+1)^3(m+2)(m+3)} + \cdots$$

where the prime indicates the derivative, and the argument of f and its derivatives is $1 \Big/ F\left(\dfrac{i}{m+1}\right)$. Thus in principle we could obtain the roots r_m and therefore the maximizing value of m for any differentiable component failure distribution $F(t)$, although the computation would generally be tedious.

CHAPTER 7
Qualitative Relationships for Multicomponent Structures

1. INTRODUCTION

In this chapter we consider redundancy models and their generalizations in which qualitative relationships between structure reliability and component reliabilities are obtained.

The problem of constructing reliable basic components by appropriate redundant use of relatively unreliable components, discussed in Section 2, was first studied by von Neumann (1956). He shows how to combine a number of unreliable "Sheffer stroke" organs to obtain an element which acts like a Sheffer stroke organ of higher reliability. (A Sheffer stroke is a device with two binary inputs and one binary output which performs a basic logical operation on two variables "not A and not B"; all logical operations can be generated from this basic one.) Moore and Shannon (1956), inspired by the von Neumann paper, carry out an elegant analysis for relay circuits in which they show that by the proper incorporation of redundancy, arbitrarily reliable circuits can be constructed from arbitrarily unreliable relays. Bounds are obtained on the number of elements required to achieve a specified reliability. Moore and Shannon also prove the probability of closure of a relay network plotted as a function of probability of closure of a single relay crosses the diagonal at most once, that is, is S-shaped. This property is crucially important in their construction of relay circuits of arbitrarily high reliability.

Birnbaum, Esary, and Saunders (1961) generalize the concepts and extend some of the results of Moore and Shannon (including the S-shapedness property) to the large natural class of structures having the property that replacing failed components by functioning components cannot cause a functioning structure to fail (called "coherent" structures by Birnbaum, Esary, and Saunders; we prefer the term "monotonic" structures for reasons which will be apparent). In Section 3 we survey their work, and also a recent generalization by Esary and Proschan (1962, 1963a) to the case of monotonic structures consisting of components of differing reliabilities. Convenient bounds are obtained on structure reliability by approximating the original arbitrary structure first by a parallel-series structure, then by a series-parallel structure. In Section 4 is presented a generalization of the S-shapedness property to monotonic structures consisting of components of differing reliabilities. In Section 5 the important class of structures which function if and only if k-out-of-n elements function is discussed. These so-called "k-out-of-n" structures are, in a certain sense, optimum in the class of monotonic structures. They include series, parallel, and "fail-safe" structures. Their mathematical tractability makes it possible to obtain interesting and useful results concerning their reliability, including Chebyshev-type bounds under minimal assumptions on component failure distributions.

Finally, in Section 6 we study the relationship between the failure rate of a structure and the failure rates of the components. We show, for example, that k-out-of-n structures and structures formed from their compositions have increasing failure rate if the individual components independently have a common failure distribution with increasing failure rate. For systems of nonidentical components we obtain bounds on structure failure rate in terms of component failure rates. For the case of k-out-of-n structures and their compositions the bounds are shown to possess certain monotonicity properties.

2. ACHIEVING RELIABLE RELAY CIRCUITS

The Moore-Shannon papers (1956) represent a complete formal solution, elegantly attained and clearly exposed, of the problem of constructing relay circuits of arbitrarily high reliability from like components of any given degree of unreliability. The exact model treated may be summarized as follows.

Suppose like idealized relays are subject to two kinds of failure: failure to close and failure to open. Similarly, circuits constructed from these relays are subject to two kinds of failure: failure to close, that is, no closed path is achieved from input wire to output wire when the circuit is commanded to close, and failure to open, that is, a closed path exists from input wire to output wire even though the circuit is commanded to open. Each relay operates independently of the others with probability p_1 (preferably small) of closing when commanded to open (nonenergized) and probability p_2 (preferably large) of closing when commanded to close (energized). These probabilities do not vary with time. The basic problem is to achieve a probability, call it $h(p_1)$, sufficiently small, of the circuit being closed when commanded to be open, and a probability, call it $h(p_2)$, sufficiently large, of the circuit being closed when commanded to be closed, by appropriate arrangement of a sufficient number of primitive units.

It soon becomes clear that the analysis may be conducted more simply by studying $h(p)$, the probability of circuit closure, as a function of p,

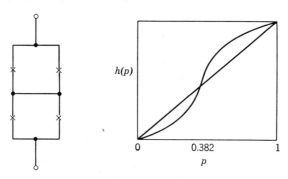

Fig. 2.1. A series-parallel circuit.

the common probability of individual relay closure. For p small (corresponding to a nonenergized relay) we would hope to achieve a value $h(p) < p$, whereas for p large (corresponding to an energized relay) we would hope to achieve a value $h(p) > p$. Thus the relay circuit represents an improvement in reliability over the individual relay when the function $h(p)$ is S-shaped. For example, consider the network and associated function shown in Figure 2.1.

Note that the associated function

$$h(p) = [1 - (1 - p)^2]^2 = 4p^2 - 4p^3 + p^4$$

does have the S-shape, with $h(p) < p$ for $p < .382$ and $h(p) > p$ for $p > .382$. Thus in this case if the probability of failure of a relay to open is $< .382$ and the probability of failure of a relay to close is $< .618$, the series-parallel network will represent an improvement over a single relay.

Some general properties may be derived for $h(p)$. Note first that

$$h(p) = \sum_{i=0}^{n} A_i p^i (1 - p)^{n-i}, \qquad (2.1)$$

where n is the total number of relays in the network, and A_i is the number of ways we can select a subset of i relays in the network such that if these

MULTICOMPONENT STRUCTURES 199

relays are closed and the remaining relays open, the network will be closed.

An important and interesting general result concerning the shape of the $h(p)$ curve is the following.

THEOREM 2.1 (Moore and Shannon). Let $h(p) \not\equiv p$ be an arbitrary network function. If $h(p_0) = p_0$ for some $0 < p_0 < 1$, then $h(p) < p$ for $0 \le p < p_0$, and $h(p) > p$ for $p_0 < p \le 1$.

This means that for any network the function $h(p)$ crosses the diagonal line of slope 1 at most once in the interval (0, 1), the crossing occurring from below.

The proof depends on first proving the differential inequality

$$\frac{h'(p)}{[1 - h(p)]h(p)} > \frac{1}{(1 - p)p}, \quad 0 < p < 1. \qquad (2.2)$$

Theorem 2.1 then follows since one solution of the corresponding equality

$$\frac{h'(p)}{[1 - h(p)]h(p)} = \frac{1}{(1 - p)p}$$

is $h(p) = p$; that is, the inequality implies that at a point of crossing of the straight line of slope 1, the function $h(p)$ must have the greater slope, limiting the number of crossings to one at most.

We shall not present the details of the Moore-Shannon proof, but instead in Section 4 we shall derive the result for a more general class of networks. In the process we shall derive several new results of a similar type.

Another upper bound on the $h(p)$ function may be obtained for a network of n relays. Consider not just the physically realizable $h(p)$ functions but all functions of the type

$$h(p) = \sum_{i=0}^{n} A_i p^i (1 - p)^{n-i}$$

where $A_i \le \binom{n}{i}$ with $i = 0, 1, 2, \ldots, n$. (This last inequality simply reflects the fact that there are at most $\binom{n}{i}$ ways in which i relays can be closed with the remaining relays open such that the network is closed.) Moore and Shannon call a function $h_Q(p)$ a *quorum* function if for some $0 \le s \le n$

$$h_Q(p) = A_s p^s (1 - p)^{n-s} + \sum_{i=s+1}^{n} \binom{n}{i} p^i (1 - p)^{n-i}, \qquad (2.3)$$

where $A_s \le \binom{n}{s}$. A quorum function corresponds to a network which is closed if for some $0 \leq s \leq n$ more than s of the relays are closed, and which is open if for some $0 \leq s \leq n$ fewer than s of the relays are closed. It may then be shown that for networks of a given number of elements, say n, quorum functions have the greatest sharpening effects on component reliability (i.e., reduce probability of closure for small component probability of closure and increase probability of closure for high component probability of closure):

THEOREM 2.2 (Moore and Shannon). Let $h_Q(p)$ be a quorum function for a network of n elements, and $h(p)$ the function for any other network

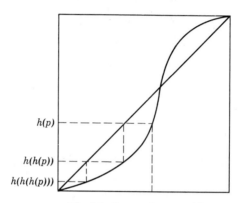

Fig. 2.2. Repeated composition.

of n elements. If for some p_0, $0 < p_0 < 1$, $h_Q(p_0) = h(p_0)$, then

$$h_Q(p) < h(p), \quad 0 \leq p < p_0$$

whereas $\quad h_Q(p) > h(p), \quad p_0 < p \leq 1.$

We shall not present the proof as given by Moore and Shannon, but rather in Section 5 we show how to obtain this result more simply for a more general class of networks.

Networks may be combined in various ways.

1. If network 1 having function $h_1(p)$ is connected in series with network 2 having function $h_2(p)$, the resulting network will have function $h_1(p)h_2(p)$.

2. If the two networks are combined in parallel, the resulting network will have function $1 - [1 - h_1(p)][1 - h_2(p)]$.

3. If network 1 is composed of elements each of which is a copy of network 2, the resulting function is

$$h(p) = h_1(h_2(p)).$$

If $h_1 \equiv h_2 \equiv h$ and this process of "composition" is repeatedly performed we obtain

$$h^{(n)}(p) = h(h(h \ldots h(p) \ldots)).$$

This may be represented graphically as in Figure 2.2.

Note that repeated composition of the function $h(p)$ should lead to a step function for $\lim_{n\to\infty} h^{(n)}(p)$, since ordinates to the left of p_0 (where $h(p_0) = p_0$) approach 0, ordinates to the right of p_0 approach 1, whereas $h^{(n)}(p)$ remains equal to p_0 for each value of n.

This argument can be made rigorous as follows. Suppose $h(p_0) = p_0$ for $0 < p_0 < 1$; then for $p < p_0$, $h(p) < p$ implies $h^{(n)}(p) < h^{(n-1)}(p) < \ldots < p$. Therefore $h^{(n)}(p)$ decreases to a limit, say L, for $p < p_0$. Since $h(p)$, a polynomial, is continuous in p we have

$$h(L) = h\left[\lim_{n\to\infty} h^{(n)}(p)\right] = \lim_{n\to\infty} h[h^{(n)}(p)] = \lim_{n\to\infty} h^{(n)}(p) = L.$$

Therefore $h(L) = L$, and L must be zero by Theorem 2.1 since $h(p)$ can cross the diagonal at most once in the open interval $(0, 1)$. A similar argument proves that $\lim_{n\to\infty} h^{(n)}(p) = 1$ for $p > p_0$.

As indicated earlier, the central problem in the Moore-Shannon paper is the construction of circuits of arbitrarily high reliability from arbitrarily unreliable relays. The key result obtained is that given $\delta > 0$, $0 < a < c < 1$, we may construct a network such that $h(a) < \delta$, $h(c) > 1 - \delta$, using at most N contacts, where

$$N = 81\left[\frac{\log \frac{c-a}{4}}{\log d}\right]\left(\frac{1}{c-a}\right)^{\frac{\log 9}{\log 3/2}}\left(\frac{\log \sqrt{8\delta}}{\log \sqrt{8}}\right)^2$$

and

$$d = \max\left(\frac{a+c}{2}, 1 - \frac{a+c}{2}\right).$$

This means that from relays which fail to open with probability a and fail to close with probability c, we may construct a network which will reduce these unreliabilities to less than δ, an arbitrarily small positive number. The number of relays required is bounded by the value N above.

In addition, it is possible to obtain a *lower* bound on the number of relays required to achieve a specified level in reliability. For a network satisfying

$$h(a) \leq \delta_1,$$
$$h(c) \geq 1 - \delta_2$$

the number of relays required is at least

$$\frac{\log \delta_1}{\log a} \cdot \frac{\log \delta_2}{\log (1 - c)}$$

(Moore and Shannon, 1956).

Although the results have been derived on the assumption that the probability of a relay closing is constant over time, the analysis may be extended to cover a different situation. Consider a network of n like relays each of which has a probability $F(t)$ of closing by time t, independent of the others. Then the probability of the network being closed by time t is simply $h(F(t))$. If the network is scheduled to close at a specified time t_0, we would want $h(F(t))$ to be small for $t < t_0$ and large for $t > t_0$. In this context we may readily interpret Theorems 2.1 and 2.2 of this chapter. Theorem 2.1 shows that a single component network is either not comparable to a given network or yields less precise timing. Theorem 2.2 shows that a quorum network gives the most precise timing among networks of a given size.

3. MONOTONIC STRUCTURES

Structure function

In the present section we survey research in redundancy leading to general qualitative relationships true for a broad class of structures. This class is systematically studied by Birnbaum, Esary, and Saunders (1961). We present here some of their concepts and results, along with some additional results obtained by J. D. Esary and F. Proschan (1962 and 1963a).

Consider structures capable of two states of performance, either complete success in accomplishing an assigned function or complete failure to function. Similarly, the components from which the structures are constructed are assumed capable of only the same two states of performance. The performance of the structure is represented by an indicator φ which is given the value 1 when the system functions, 0 when the system fails. The performance of each of the n components in the structure is similarly represented by an indicator x_i which takes the value 1 if the ith component functions, 0 if the ith component fails ($i = 1, 2, \ldots, n$).

It is assumed that the performance of a structure depends deterministically on the performance of the components in the structure, that is, that φ is a function of $\mathbf{x} = (x_1, x_2, \ldots, x_n)$; $\varphi(\mathbf{x})$ is called the *structure function* of the structure.

For structures in which each component if functioning contributes to the functioning of the structure, certain hypotheses appear intuitively

acceptable:
 (i) $\varphi(\mathbf{1}) = 1$, where $\mathbf{1} = (1, 1, \ldots, 1)$,
 (ii) $\varphi(\mathbf{0}) = 0$, where $\mathbf{0} = (0, 0, \ldots, 0)$,
 (iii) $\varphi(\mathbf{x}) \geq \varphi(\mathbf{y})$ whenever $x_i \geq y_i$, $i = 1, 2, \ldots, n$.

Hypothesis (i) states that if all the components function, the structure functions. Hypothesis (ii) states that if all the components fail, the structure fails. Finally, hypothesis (iii) states that functioning components do not interfere with the functioning of the structure. Structures satisfying (i), (ii), and (iii) are called *coherent* in Birnbaum, Esary, and Saunders (1961); we prefer the term *monotonic* since such structures are characterized by a monotonic structure function which is 0 at $\mathbf{0}$, 1 at $\mathbf{1}$.

Some examples of monotonic structures follow.

Example 1. A simple two-terminal network

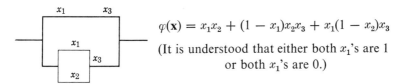

$\varphi(\mathbf{x}) = x_1 x_2 + x_1(1 - x_2) x_3$

Example 2. A two-terminal network with replication of components

$\varphi(\mathbf{x}) = x_1 x_2 + (1 - x_1) x_2 x_3 + x_1(1 - x_2) x_3$

(It is understood that either both x_1's are 1 or both x_1's are 0.)

Example 3. The 2-out-of-3 structure (the structure functions if at least two components function)

$\varphi(\mathbf{x}) = x_1 x_2 (1 - x_3) + x_1(1 - x_2) x_3 + (1 - x_1) x_2 x_3 + x_1 x_2 x_3$

Paths and cuts

Two-terminal networks with binary components are monotonic structures. For such networks, the notions of path and cut have physical significance. The notion extends quite naturally to binary structures of binary components. Each specification of component performances $\mathbf{x} = (x_1, x_2, \ldots, x_n)$ determines a partition A, B of the set of components according to
$$A = \{i : x_i = 1\}, \qquad B = \{i : x_i = 0\}.$$
If $\varphi(\mathbf{x}) = 1$ and $\varphi(\mathbf{y}) = 0$ for any $\mathbf{y} \leq \mathbf{x}$ but $\neq \mathbf{x}$, then A is a *path* of the system. If $\varphi(\mathbf{x}) = 0$ and $\varphi(\mathbf{y}) = 1$ for any $\mathbf{y} \geq \mathbf{x}$ but $\neq \mathbf{x}$, then B is called a *cut*. For monotonic structures a path is a minimal set of components which by functioning ensure the functioning of the structure. A cut is a

minimal set of components which by failing guarantee the failure of the structure. The *size* of a path (cut) is the number of components in the path (cut).

With each path A_j, $j = 1, 2, \ldots, r$ say, we may associate a binary function

$$\alpha_j(\mathbf{x}) = \prod_{i \in A_j} x_i \tag{3.1}$$

which takes the value of 1 if all components in the path function and 0 otherwise. Clearly, α_j is the structure function of a structure in which the components of the jth path act in series. Similarly, with each cut B_k, $k = 1, 2, \ldots, s$ say, we may associate a binary function

$$\beta_k(\mathbf{x}) = 1 - \prod_{i \in B_k} (1 - x_i) \tag{3.2}$$

which takes the value 0 if all components in the kth cut fail and is 1 otherwise. Thus β_k is the structure function of a structure in which the components of the kth cut act in parallel. It is possible (Birnbaum, Esary, and Saunders, 1961) to represent the structure function φ of a coherent structure in terms of either $\alpha_1, \alpha_2, \ldots, \alpha_r$ or $\beta_1, \beta_2, \ldots, \beta_s$ by

$$\varphi(\mathbf{x}) = 1 - \prod_{j=1}^{r} [1 - \alpha_j(\mathbf{x})], \tag{3.3}$$

or by

$$\varphi(\mathbf{x}) = \prod_{k=1}^{s} \beta_k(\mathbf{x}), \tag{3.4}$$

that is, as a structure for which the path structures act in parallel or as a structure for which the cut structures act in series. Note that the definitions of path, cut, and size used here differ, but only technically, from those used in Birnbaum, Esary, and Saunders (1961) and in J. D. Esary and F. Proschan (1962a).

Example 4. Bridge

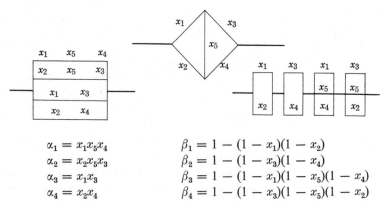

$\alpha_1 = x_1 x_5 x_4$ $\beta_1 = 1 - (1 - x_1)(1 - x_2)$
$\alpha_2 = x_2 x_5 x_3$ $\beta_2 = 1 - (1 - x_3)(1 - x_4)$
$\alpha_3 = x_1 x_3$ $\beta_3 = 1 - (1 - x_1)(1 - x_5)(1 - x_4)$
$\alpha_4 = x_2 x_4$ $\beta_4 = 1 - (1 - x_3)(1 - x_5)(1 - x_2)$

Reliability

Now assume a probability distribution for the performance of the components. In particular, let

$$p_i = P[X_i = 1] = EX_i$$

be the *reliability* of the ith component, where X_i is the binary random variable designating the state of component i. The structure function $\varphi(\mathbf{x})$ of a structure now becomes a binary random variable with a probability distribution determined by the probability distribution of $X_1, X_2, \ldots, X_n$. We call

$$h = P[\varphi(\mathbf{X}) = 1] = E\varphi(\mathbf{X})$$

the *reliability* of the structure. When components perform independently, as we assume for the rest of the chapter, the distribution of $X_1, X_2, \ldots, X_n$ is determined from $p_1, p_2, \ldots, p_n$, and so we may write

$$h = h(\mathbf{p}),$$

where $\mathbf{p} = (p_1, p_2, \ldots, p_n)$; $h(\mathbf{p})$ is the *reliability function* of the structure. If $p_1 = p_2 = \cdots = p_n = p$, we shall write $h(p)$.

Note that for a monotonic structure with independent components

$$h(\mathbf{0}) = E[\varphi(\mathbf{X}) \,|\, p_1 = 0, \ldots, p_n = 0] = \varphi(\mathbf{0}) = 0$$

and $\quad h(\mathbf{1}) = E[\varphi(\mathbf{X}) \,|\, p_1 = 1, \ldots, p_n = 1] = \varphi(\mathbf{1}) = 1.$

Moreover, $h(\mathbf{p})$ is increasing in each p_i as we now show. For $h(\mathbf{p})$ of order n, write

$$\begin{aligned} h(\mathbf{p}) &= E\varphi(X_1, \ldots, X_n) = p_n E\varphi(X_1, \ldots, X_{n-1}, 1) \\ &\quad + q_n E\varphi(X_1, \ldots, X_{n-1}, 0) \\ &= E\varphi(X_1, \ldots, X_{n-1}, 0) + p_n E[\varphi(X_1, \ldots, X_{n-1}, 1) \\ &\quad - \varphi(X_1, \ldots, X_{n-1}, 0)]. \end{aligned}$$

Since $\quad \varphi(X_1, \ldots, X_{n-1}, 1) - \varphi(X_1, \ldots, X_{n-1}, 0) \geq 0$

(φ being a monotonic structure), the assertion follows for $h(\mathbf{p})$ as a function of p_n; similarly the assertion follows for $h(\mathbf{p})$ as a function of $p_1, \ldots, p_{n-1}$.

Using this fact we may obtain the following lower bound on system reliability for monotonic structures.

THEOREM 3.1. *Let F_i, IFR with mean μ_i, $i = 1, \ldots, n$, be the failure distributions of the components of a monotonic structure. Then for*

$t < \min(\mu_1, \ldots, \mu_n)$, the system reliability $h(\bar{F}_1(t), \ldots, \bar{F}_n(t)) \geq h(e^{-t/\mu_1}, \ldots, e^{-t/\mu_n})$, the corresponding system reliability when the components are exponentially distributed with the same means.

Proof. First recall that $h(p_1, \ldots, p_n)$ is monotonic increasing in each argument. Next note from Theorem 4.4 of Chapter 2 that for

$$t < \min(\mu_1, \ldots, \mu_n), \quad \bar{F}_i(t) \geq e^{-t/\mu_i} \quad (i = 1, \ldots, n).$$

It follows that for

$$t < \min(\mu_1, \ldots, \mu_n), h(\bar{F}_1(t), \ldots, \bar{F}_n(t)) \geq h(e^{-t/\mu_1}, \ldots, e^{-t/\mu_n}). \parallel$$

Note that Theorem 3.1 generalizes the corresponding result for series and parallel structures following Theorem 4.4 of Chapter 2. In a similar fashion we may prove

THEOREM 3.2. *Let F, IFR with mean μ but not identically exponential, be the common failure distribution of the components of a monotonic structure having reliability function $h(p)$. Let $t_0 > \mu$ be the point of crossing of $\bar{F}(t)$ and $e^{-t/\mu}$. Then system survival probability $h(\bar{F}(t))$ using IFR components crosses system survival probability $h(e^{-t/\mu})$ using exponential components once only, at t_0, and from above.*

Path and cut bounds on structure reliability

In J. D. Esary and F. Proschan (1963a) the following bounds are obtained on the reliability of monotonic structures of independent components:

$$\prod_{k=1}^{s} P[\beta_k(\mathbf{X}) = 1] \leq P[\varphi(\mathbf{X}) = 1] \leq 1 - \prod_{j=1}^{r} \{1 - P[\alpha_j(\mathbf{X}) = 1]\}, \quad (3.5)$$

where α_j $(j = 1, 2, \ldots, r)$ and β_k $(k = 1, 2, \ldots, s)$ are the path and cut structure functions as defined by (3.1) and (3.2). The probabilities in the bounding expressions may be evaluated by the usual series and parallel formulas:

$$P[\alpha_j(\mathbf{X}) = 1] = \prod_{i \in A_j} p_i \quad (3.6)$$

$$P[\beta_k(\mathbf{X}) = 1] = 1 - \prod_{i \in B_k}(1 - p_i). \quad (3.7)$$

These bounds on the reliability function $h(\mathbf{p}) = E\varphi(\mathbf{X})$ of the structure are, in a sense, the reliability analogues of the representations (3.3) and (3.4) for the structure function.

The intuitive motivation for them is quite simple. For example, the representation (3.3) corresponds to arranging the components of each path to act in series, and arranging the resulting path substructures to act in parallel. Of course the same component may appear in more than one path so that in considering the structure from a physical point of view, it is necessary to suppose some mechanism which causes replications of the same component to either all function or all fail. The right-hand expression in (3.5) is the structure reliability that would result if each linked replication of a component were replaced by an independently performing version of the component, a process which should increase structure reliability.

The theoretical interest of the bounds is that they offer the means in reliability studies of approximating structures having complex component arrangements with structures having only series and parallel arrangements. The practical interest of the bounds is that they can be useful for structures having reliability functions tedious to evaluate exactly, but whose paths and cuts can be determined by inspection. Such structures are quite numerous.

The proof of the validity of the bounds (3.5) is given in detail in J. D. Esary and F. Proschan (1963a). We shall sketch the main ideas in the proof. A key tool used in obtaining this result as well as other results in structural reliability is the specialized version of the Chebyshev inequality (G. H. Hardy, J. E. Littlewood, and G. Pólya, 1952) stating that the covariance of two increasing functions of independent binary random variables is nonnegative.

THEOREM 3.3. Let $X_1, X_2, \ldots, X_n$ be independent binary random variables. Let $f_j(\mathbf{X})$ ($j = 1, 2$) be increasing (decreasing) functions. Then cov $[f_1(\mathbf{X}), f_2(\mathbf{X})] \geq 0$.

The proof of Theorem 3.3 is obtained by first demonstrating that for any $f(\mathbf{X})$:

$$Ef(\mathbf{X}) = p_i Ef(1_i, \mathbf{X}) + q_i Ef(0_i, \mathbf{X}), \tag{3.8}$$

where $\quad f(1_i, \mathbf{X}) = f(X_1, \ldots, X_{i-1}, 1, X_{i+1}, \ldots, X_n)$

and $\quad f(0_i, \mathbf{X}) = f(X_1, \ldots, X_{i-1}, 0, X_{i+1}, \ldots, X_n)$.

It readily follows that for any functions $f_j(\mathbf{X})$ ($j = 1, 2$) and any X_i selected from independent binary random $X_1, \ldots, X_n$:

$$\text{cov}\,[f_1(\mathbf{X}), f_2(\mathbf{X})] = p_i \,\text{cov}\,[f_1(1_i, \mathbf{X}), f_2(1_i, \mathbf{X})] + q_i \,\text{cov}\,[f_1(0_i, \mathbf{X}), f_2(0_i, \mathbf{X})]$$
$$+ p_i q_i E[f_1(1_i, \mathbf{X}) - f_1(0_i, \mathbf{X})] E[f_2(1_i, \mathbf{X}) - f_2(0_i, \mathbf{X})]. \tag{3.9}$$

Finally, by induction on the order n of the structure it is easy to verify from (3.9) that the conclusion of Theorem 3.1 holds.

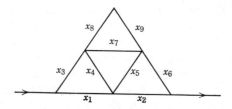

Fig. 3.1 Two-terminal network.

Using Theorem 3.3 we may readily prove the inequalities (3.5). From (3.4), since each $\beta_k(\mathbf{x})$ is an increasing function of $x_1, \ldots, x_n$, we have

$$E\varphi(\mathbf{X}) = E\prod_{k=1}^{s}\beta_k(\mathbf{X}) \geq \prod_{k=1}^{s}E\beta_k(\mathbf{X}) = \prod_{k=1}^{s}P[\beta_k(\mathbf{X}) = 1],$$

the inequality resulting from repeated applications of Theorem 3.1. Similarly from (3.3)

$$E[1 - \varphi(\mathbf{X})] = E\prod_{j=1}^{r}[1 - \alpha_j(\mathbf{X})] \geq \prod_{j=1}^{r}E[1 - \alpha_j(\mathbf{X})] = \prod_{j=1}^{r}P[1 - \alpha_j(\mathbf{X})].$$

As an example consider the two-terminal network shown in Figure 3.1.

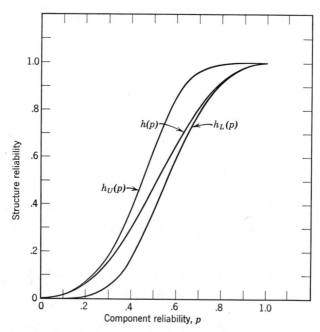

Fig. 3.2. Bounds on structure reliability for a two-terminal network.

The paths are readily seen to be

12	156	1476	14896	342
3456	376	3752	3896	38952.

Similarly, the cuts are

13	1478	1479	1456
2543	2578	2579	26.

From these paths and cuts we immediately obtain in the case of like components, using (3.5):

$$h_L(p) \le h(p) \le h_U(p),$$

where $h_L(p) = (1 - q^2)^2(1 - q^4)^6$,

and $h_U(p) = 1 - (1 - p^2)(1 - p^3)^3(1 - p^4)^4(1 - p^5)^2$.

Figure 3.2 shows graphically $h(p)$, $h_L(p)$, and $h_U(p)$.

4. S-SHAPED RELIABILITY FUNCTIONS FOR MONOTONIC STRUCTURES

We have seen in Section 2 that for relay networks an S-shaped network function represents an improvement over the network function of a single relay. We have also seen in Theorem 2.1 that the network function of any relay network, if it crosses the diagonal, does so once exactly and from below, i.e., is S-shaped. This result is extended in Z. W. Birnbaum, J. D. Esary, and S. C. Saunders (1961) to the reliability functions of monotonic structures of like components. Finally, in J. D. Esary and F. Proschan (1963a) an extension of sorts of this property is shown to hold for the reliability function of monotonic structures of components not necessarily alike.

It should be noted, however, that the S-shapedness of the probability of closure function of a relay network has a different interpretation than does the S-shapedness of the reliability function of a monotonic structure. For a relay network, a low probability p of the relay closing corresponds to a command to open (see Section 2); we thus would like to obtain an even lower probability $h(p)$ that the network closes for small p. On the other hand, in the case of the reliability of a monotonic structure, we almost never wish to achieve an even lower reliability for the structure than for the component, even for small p. It is useful, however, to know that for some p_0, structure reliability is larger than component reliability for $p > p_0$. Thus we are interested in the S-shapedness of both the reliability function and the relay network function.

To establish the S-shapedness of the reliability functions of monotonic structures, we first prove:

LEMMA 4.1. Let $\varphi(\mathbf{x})$ be the structure function of a monotonic structure of components having respective reliabilities $p_1, p_2, \ldots, p_n$. Then

$$\text{cov}\left[\varphi(\mathbf{X}), \sum_{i=1}^{n} X_i\right] \geq \text{var } \varphi(\mathbf{X}). \tag{4.1}$$

Proof. It is sufficient to prove that

$$\text{cov}\left[\varphi(\mathbf{X}), \sum_{1}^{n} X_i - \varphi(\mathbf{X})\right] \geq 0. \tag{4.2}$$

But the last inequality follows from Theorem 3.3 since both $\varphi(\mathbf{X})$ and $\sum_{1}^{n} X_i - \varphi(\mathbf{X})$ are increasing functions. ‖

Actually, if the monotonic structure has at least two essential components, strict inequality holds in (4.1), as shown in J. D. Esary and F. Proschan (1963a). (Component i is essential if $\varphi(1_i, \mathbf{x}) \not\equiv \varphi(0_i, \mathbf{x})$ for all $\mathbf{x}$.)

Suppose now we choose $f_1(\mathbf{X}) = \varphi(\mathbf{X})$, $f_2(\mathbf{X}) = X_i$ in (3.9). We get

$$\text{cov }[\varphi(\mathbf{X}), X_i] = p_i q_i E[\varphi(1_i, \mathbf{X}) - \varphi(0_i, \mathbf{X})]. \tag{4.3}$$

But since

$$h(\mathbf{p}) = p_i E\varphi(1_i, \mathbf{X}) + q_i E\varphi(0_i, \mathbf{X}),$$

differentiation yields

$$\frac{\partial h(\mathbf{p})}{\partial p_i} = E\varphi(1_i, \mathbf{X}) - E\varphi(0_i, \mathbf{X}). \tag{4.4}$$

Substitution into (4.3) gives

$$\text{cov }[\varphi(\mathbf{X}), X_i] = p_i q_i \frac{\partial h}{\partial p_i}.$$

By summing over i we get

$$\text{cov}\left[\varphi(\mathbf{X}), \sum_{i=1}^{n} X_i\right] = \sum_{i=1}^{n} p_i q_i \frac{\partial h}{\partial p_i}. \tag{4.5}$$

Since $\varphi(\mathbf{X})$ is a binomial variable, we know that

$$\text{var } \varphi(\mathbf{X}) = h(\mathbf{p})[1 - h(\mathbf{p})]. \tag{4.6}$$

Substituting (4.5) and (4.6) into (4.1) gives us a generalization of the Moore-Shannon inequality of Equation (2.2):

THEOREM 4.2. Let $h(\mathbf{p})$ be the reliability function of a monotonic structure of independent components having at least two essential components. Then

(i) $$\sum_{i=1}^{n} p_i q_i \frac{\partial h(\mathbf{p})}{\partial p_i} > h(\mathbf{p})[1 - h(\mathbf{p})]. \qquad (4.7)$$

(ii) If $p_1 = p_2 = \cdots = p_n = p$ and $h(p_0) = p_0$ for some $0 < p_0 < 1$, then $h(p) < p$ for $0 \leq p < p_0$, whereas $h(p) > p$ for $p_0 < p < 1$.

In the special case $p_1 = p_2 = \cdots = p_n = p$, (ii) is the Moore-Shannon inequality, Equation (2.2), obtained for two-terminal networks in E. F. Moore and C. E. Shannon (1956) and for monotonic structures in Birnbaum, Esary, and Saunders (1961). As we remarked in Section 2, Equation (2.2) implies that at an intersection of $h(p)$ with the diagonal, $h'(p) > 1$, so that $h(p)$ can intersect the diagonal at most once, and from below, that is, $h(p)$ is S-shaped.

In Moore and Shannon (1956), where only the case of like components is considered, the family of curves

$$\frac{f_c(p)}{1 - f_c(p)} = c \frac{p}{1 - p}, \quad c > 0, \quad 0 \leq p \leq 1, \qquad (4.8)$$

is introduced. For this family $f_c(0) = 0$, $f_c(1) = 1$, and f_c is a monotone increasing function of p. The "diagonal" $f_c(p) = p$ occurs when $c = 1$. If $f_c(p)$ is interpreted as the reliability of a structure with components of reliability p, the odds $f_c/(1 - f_c)$ that the structure will function are proportional to the odds $p/(1 - p)$ that a component will function. Taking logarithms on both sides of (4.8) and differentiating, we verify that the curves satisfy

$$p(1 - p) \frac{df_c(p)}{dp} = f_c(p)[1 - f_c(p)].$$

Applying the Moore-Shannon inequality (2.2), we see that if $h(p_o) = f_c(p_o)$ for some p_o $(0 < p_o < 1)$, then at p_o

$$p_o(1 - p_o) \frac{dh(p_o)}{dp} > h(p_o)[1 - h(p_o)] = f_c(p_o)[1 - f_c(p_o)]$$

$$= p_o(1 - p_o) \frac{df_c(p_o)}{dp},$$

so that
$$\frac{dh(p_o)}{dp} > \frac{df_c(p_o)}{dp}.$$

Thus the reliability function of a monotonic structure with identical components having at least two essential components can cross an f_c curve at most once, and only from below. For a structure with just one essential component (that is, $h(p) = p$), the reliability function $h(p)$ coincides with the "diagonal" curve f_1.

By a simple auxiliary argument it is possible to determine which of the proportional odds curves actually are crossed by the reliability function of a monotonic structure. For this we evaluate $dh(p)/dp$ at $p = 0$ and at $p = 1$. From (4.4)

$$\frac{dh(p)}{dp} = \sum_{i=1}^{n} \frac{\partial h(\mathbf{p})}{\partial p_i} \frac{dp_i}{dp} = \sum_{i=1}^{n} E[\varphi(1_i, \mathbf{X}) - \varphi(0_i, \mathbf{X})],$$

so that
$$\left.\frac{dh}{dp}\right|_{p=0} = \sum_{i=1}^{n} [\varphi(1_i, \mathbf{0}) - \varphi(0_i, \mathbf{0})] = \sum_{i=1}^{n} \varphi(1_i, \mathbf{0}),$$

since

$$E[\varphi(1_i, \mathbf{X}) \mid p = 0] = \varphi(1_i, \mathbf{0}) \quad \text{and} \quad E[\varphi(0_i, \mathbf{X}) \mid p = 0] = \varphi(0_i, \mathbf{0}) = 0.$$

Recall that
$$\varphi(1_i, \mathbf{X}) = \varphi(X_1, X_2, \ldots, X_{i-1}, 1, X_{i+1}, \ldots, X_n)$$
$$\varphi(0_i, \mathbf{X}) = \varphi(X_1, X_2, \ldots, X_{i-1}, 0, X_{i+1}, \ldots, X_n).$$

Therefore $\varphi(1_i, \mathbf{0})$ is 1 if and only if the ith component is a path of size 1; otherwise $\varphi(1_i, \mathbf{0}) = 0$. Thus

$$\left.\frac{dh}{dp}\right|_{p=0} = \text{number of paths of size 1}.$$

Similarly

$$\left.\frac{dh}{dp}\right|_{p=1} = \sum_{i=1}^{n} \{\varphi(1_i, \mathbf{1}) - \varphi(0_i, \mathbf{1})\} = \sum_{i=1}^{n} \{1 - \varphi(0_i, \mathbf{1})\}$$

or equals the number of cuts of size 1.

For the f_c curves,
$$\left.\frac{df_c}{dp}\right|_{p=0} = c, \quad \left.\frac{df_c}{dp}\right|_{p=1} = \frac{1}{c}.$$

Comparison of derivatives at $p = 0$ and $p = 1$ leads to the following classification; we make application of it in Section 5 to k-out-of-n structures.

TABLE 4.1†

No. of Paths of Size 1	No. of Cuts of Size 1	Slope of Reliability Curve $h(p)$ Relative to $f_c(p)$
0	0	$h'(0) = 0$, $h'(1) = 0$; $h(p)$ crosses each curve $f_c(p)$ $(0 < c < \infty)$ at some p_o $(0 < p_o < 1)$.
j $(1 \leq j \leq n)$	0	$h'(0) = j$, $h'(1) = 0$; $h(p)$ is tangent to $f_j(p)$ at $p = 0$, $h(p)$ crosses each curve $f_c(p)$ $(j < c < \infty)$ at some p_o $(0 < p_o < 1)$.
0	k $(1 \leq k \leq n)$	$h'(1) = 0$, $h'(1) = k$; $h(p)$ is tangent to $f_{1/k}(p)$ at $p = 1$, $h(p)$ crosses each curve $f_c(p)$ $(0 < c < 1/k)$ at some p $(0 < p < 1)$.
1	1	$h(p) = p = f_1(p)$; the structure has just one essential component.

† From Esary and Proschan, The reliability of coherent systems, in *Redundancy Techniques for Computing Systems*, edited by Wilcox and Mann, Spartan Books, Washington, D.C., 1962.

The term S-shapedness arises (Birnbaum, Esary, and Saunders, 1961) from the shape of the reliability function of monotonic structures having no paths or cuts of size 1, relative to the diagonal curve $f_1(p) = p$ which represents component reliability.

Next we shall use the multivariate Moore-Shannon inequality (4.7) to establish an S-shapedness property for the reliability function of monotonic structures consisting of components of differing reliabilities. In the case of unlike components, structure reliability as a function of the n individual component reliabilities constitutes an n-dimensional surface. To study the shape of the reliability function meaningfully, we consider component reliabilities to be functions of a common parameter. For example, component reliabilities, although not alike, might increase together as time passes as a result of an engineering development program; thus the component reliability $p_i(t)$ of the ith component $(i = 1, 2, \ldots, n)$, might be taken as a function of the common parameter time. Alternately, the total amount of money spent in developing the structure might constitute the common parameter determining component reliability. For such situations, the following result is of interest.

THEOREM 4.3. Let $p_i(\theta)$ satisfy for $i = 1, 2, \ldots, n$

$$\theta(1 - \theta)p_i'(\theta) \geq p_i(\theta)[1 - p_i(\theta)], \quad 0 < \theta < 1. \quad (4.9)$$

Let $h(\mathbf{p})$ be the reliability function of a monotonic structure having at least two essential components. Then

(i) $\theta(1 - \theta)\dfrac{dh(\mathbf{p}(\theta))}{d\theta} > h(\mathbf{p}(\theta))\{1 - h(\mathbf{p}(\theta))\}, \quad 0 < \theta < 1;$ (4.10)

(ii) $h[\mathbf{p}(\theta)]$ crosses the diagonal θ at most once, the crossing, if it occurs, occurring from below.

Proof. From (4.7)

$$h(\mathbf{p})[1 - h(\mathbf{p})] < \sum_{i=1}^{n} p_i q_i \frac{\partial h(\mathbf{p})}{\partial p_i} \leq \sum_{i=1}^{n} \theta(1 - \theta)p_i'(\theta)\frac{\partial h}{\partial p_i} = \theta(1 - \theta)\frac{dh}{d\theta}. \quad \|$$

Note that Theorem 4.3 states essentially that the differential inequality (4.7) is preserved under composition. [The condition (4.9) would correspond roughly to reliability increasing more and more rapidly with θ until some turning point θ_o; beyond θ_o the reliability would approach 1 more and more slowly.] In the following corollary we give conditions under which $h(\mathbf{p}(\theta))$ actually does cross the diagonal precisely once.

COROLLARY. In addition to the hypotheses of Theorem 4.3, assume that each $p_i(\theta)$ actually crosses the diagonal once and that $h(p)$, the reliability function for like components, actually crosses the diagonal exactly once. Then $h(\mathbf{p}(\theta))$ crosses the diagonal exactly once.

Proof. For θ sufficiently close to 0, each $p_i(\theta) \leq \theta$, so that $h(\mathbf{p}(\theta)) \leq h(\boldsymbol{\theta}) < \theta$. Similarly for θ sufficiently close to 1, each $p_i(\theta) \geq \theta$, so that $h(\mathbf{p}(\theta)) \geq h(\boldsymbol{\theta}) > \theta$. Thus $h(\mathbf{p}(\theta))$ crosses the diagonal at least once. Since we also know from Theorem 4.3 that $h(\mathbf{p}(\theta))$ crosses the diagonal at most once, the conclusion follows.

Let us apply Theorem 4.3 in the case in which each component reliability p_i is a function of time; that is, let $p_i(t) = 1 - F_i(t)$, where $F_i(t)$ is the distribution for the life of the ith component ($i = 1, 2, \ldots, n$). If each component failure distribution has an increasing failure rate (IFR), then $h(\mathbf{p}(t))$ exhibits certain smoothness properties.

THEOREM 4.4. Assume $h(\mathbf{p})$ is the reliability function of a monotonic structure having at least two essential components. Assume also that each component life has an IFR distribution. Then $h(\mathbf{p}(t))$ crosses each curve $c/(c + t)$ ($c > 0$) at most once, and if it does, it does so from above.

Proof. Let $t = c(1/\theta - 1)$, so that $\theta = c/(c + t)$. Then the ith component reliability is given by $p_i(\theta)$ or by $\bar{F}_i(t) = 1 - F_i(t)$. Thus if $r_i(t)$

MULTICOMPONENT STRUCTURES

is the failure rate of component i, we have

$$r_i(t) = \frac{f_i(t)}{\bar{F}_i(t)} = \frac{p_i'(\theta)\theta^2/c}{p_i(\theta)}. \tag{4.11}$$

Since F_i is IFR

$$\begin{vmatrix} f_i(x) & f_i(t) \\ \bar{F}_i(x) & \bar{F}_i(t) \end{vmatrix} \leq 0$$

for $x \leq t$. The inequality remains valid if we replace $\bar{F}_i(x)$ by 1. Integrating on x from 0 to t we obtain

$$r_i(t) \geq \frac{F_i(t)}{t};$$

combined with (4.11), this yields

$$t\frac{\theta^2}{c} p_i'(\theta) \geq p_i(\theta)[1 - p_i(\theta)].$$

Since $t\theta/c = 1 - \theta$, we have

$$\theta(1 - \theta)p_i'(\theta) \geq p_i(\theta)[1 - p_i(\theta)], \quad i = 1, 2, \ldots, n.$$

Theorem 4.3 now applies, yielding the desired result. ∥

Thus we conclude that structure reliability as a function of time crosses each member of the hyperbolic family of curves at most once, assuming each component failure distribution is IFR.

Theorem 4.3 gives a lower bound for $dh(\mathbf{p}(\theta))/d\theta$ when the $p_i(\theta)$ satisfy (4.9). We can also obtain an upper bound under a related assumption on the $p_i(\theta)$.

THEOREM 4.5. Let $p_i(\theta)$ satisfy for $i = 1, 2, \ldots, n$:

$$\theta(1 - \theta)p_i'(\theta) \leq p_i(\theta)[1 - p_i(\theta)], \quad 0 < \theta < 1. \tag{4.12}$$

Let $h(\mathbf{p})$ be the reliability function of a monotonic structure having at least two essential elements. Then

$$\theta(1 - \theta)\frac{dh(\mathbf{p}(\theta))}{d\theta} < \sqrt{h(1 - h)\sum_{i=1}^{n} p_i(\theta)[1 - p_i(\theta)]}. \tag{4.13}$$

Proof.

$$\theta(1 - \theta)\frac{dh}{d\theta} = \theta(1 - \theta)\sum_{i=1}^{n}\frac{\partial h}{\partial p_i}\frac{dp_i}{d\theta} \leq \sum_{i=1}^{n}\frac{\partial h}{\partial p_i} p_i(\theta)[1 - p_i(\theta)] \quad \text{[from (4.12)]}$$

$$= \text{cov}\left[\varphi(\mathbf{X}), \sum_{i=1}^{n} X_i\right] \text{[from (4.5)]} \leq \sqrt{\text{var } \varphi(\mathbf{X}) \cdot \text{var } \sum X_i}$$

(that is, a correlation coefficient cannot exceed 1). Since var $\varphi(\mathbf{X}) = h(1 - h)$ and var $\Sigma X_i = \sum_{1}^{n} p_i(\theta)[1 - p_i(\theta)]$, (4.13) follows. ∥

In the special case of like components, $p_i(\theta) = p$ $(i = 1, 2, \ldots, n)$, we obtain the result [Moore and Shannon, 1956, Pt. II, Equation (13)]:

$$\frac{dh(p)}{dp} \leq \sqrt{\frac{n(1-h)h}{(1-p)p}}.$$

5. k-OUT-OF-n STRUCTURES

Recall that a k-out-of-n structure is a structure of order n that functions if and only if at least k components function, where $1 \leq k \leq n$. Such structures are important because, as we shall show, among all monotonic structures of order n, they display the steepest reliability functions (see Theorem 5.2 for a precise statement). It is also mathematically convenient to use k-out-of-n structures, since quantitative results are readily obtained. Note that the special case $k = n$ corresponds to the well-known series structure whereas $k = 1$ corresponds to the parallel structure. An interesting case of practical importance is that in which $k = n - 1$; that is, failure of a single component is not sufficient to cause structure failure, but the failure of two components does cause structure failure (sometimes referred to as a "fail-safe" design).

We shall treat the case in which

(i) all components have the same reliability p,
(ii) component failures are independent events.

Under these assumptions the reliability of a k-out-of-n structure is given by

$$h(p; k, n) = \sum_{i=k}^{n} \binom{n}{i} p^i (1-p)^{n-i} \tag{5.1}$$

and by a well-known formula (A. M. Mood, 1950, p. 235) this is equal to

$$h(p; k, n) = \frac{n!}{(k-1)!(n-k)!} \int_0^p x^{k-1}(1-x)^{n-k} \, dx. \tag{5.2}$$

We know from the Moore-Shannon inequality for monotonic structure functions, Equation (2.2), that $h(p; k, n)$ crosses the diagonal at most once. If $1 < k < n$, from Table 4.1 we conclude that $h(p; k, n)$ *does* cross the diagonal exactly once. We can obtain further information concerning the shape of the reliability function by taking the second derivative; we find

$$h''(p; k, n) > 0, \quad 0 < p < \frac{k-1}{n-1}$$

and

$$h''(p; k, n) < 0, \quad \frac{k-1}{n-1} < p < 1$$

for $1 < k < n$. Thus $h(p; k, n)$ is convex in p for $0 \leq p \leq (k-1)/(n-1)$ and concave for $(k-1)/(n-1) \leq p \leq 1$.

For $1 < k < n$, the point of intersection of $h(p; k, n)$ with the diagonal is found as the root $\rho_{k,n}$ of the equation

$$h(p; k, n) = p,$$

where $h(p; k, n)$ is given by (5.1) or (5.2). Numerical values of the roots

TABLE 5.1*

SOLUTIONS $\rho_{k,n}$ OF THE EQUATION $h(p; k, n) = p$ FOR $p \in (0, 1)$

n \ k	2	3	4	5	6	7	8	9	10	11	12	13
3	.500											
4	.232											
5	.131	.500										
6	.083	.347										
7	.058	.256	.500									
8	.042	.197	.395									
9	.032	.158	.322	.500								
10	.025	.129	.268	.421								
11	.020	.108	.228	.361	.500							
12	.016	.092	.197	.314	.437							
13	.013	.080	.172	.276	.387	.500						
14	.011	.070	.152	.246	.345	.448						
15	.010	.062	.136	.220	.311	.405	.500					
16	.010	.055	.122	.199	.282	.368	.456					
17	.008	.050	.111	.181	.257	.337	.418	.500				
18	.007	.045	.101	.166	.236	.310	.385	.462				
19	.006	.041	.092	.153	.218	.286	.356	.428	.500			
20	.006	.037	.085	.141	.202	.266	.331	.398	.466			
21	.005	.034	.079	.131	.188	.247	.309	.372	.436	.500		
22	.005	.032	.073	.122	.175	.231	.289	.349	.409	.470		
23	.004	.030	.068	.114	.164	.217	.272	.328	.385	.442	.500	
24	.004	.027	.064	.107	.154	.204	.256	.309	.363	.418	.473	
25	.004	.025	.060	.101	.145	.193	.242	.292	.343	.395	.448	.500

For $k > n/2$: $\rho_{k,n} = 1 - \rho_{n+1-k,n}$.

* From Birnbaum, Esary, and Saunders, Multi-component systems and structures and their reliability, *Technometrics*, V. 3, No. 1, February 1961.

$\rho_{k,n}$ are tabulated in Table 5.1 for $n = 2(1)25$ and $k = 2(1)n - 1$; note that

$$\rho_{k,n} = 1 - \rho_{n+1-k,n}.$$

Table 5.1 is taken from Birnbaum, Esary, and Saunders (1961). Table 5.1 shows at a glance for given k and n whether a k-out-of-n structure is more or less reliable than a single component of specified reliability p. If the actual value of the structure reliability $h(p; k, n)$ is required, it can be obtained from National Bureau of Standards (1950), where the expression (5.1) is tabulated.

We now proceed to prove that k-out-of-n structures have the greatest sharpening effect on individual component reliability among structures of order n. First we need a basic lemma which enables us to compare certain pairs of structures of the same order. We let A_i equal the number of different ways in which exactly i components function and at the same time the structure functions.

LEMMA 5.1. Let $h^{(j)}(p) = \sum_{i=0}^{n} A_i^{(j)} p^i q^{n-i}$, $j = 1, 2$, be the functions of two structures of order n. Suppose there exists an integer k ($1 \leq k \leq n$) such that

$$A_i^{(1)} \leq A_i^{(2)}, \quad i = 0, 1, \ldots, k$$
$$A_i^{(1)} \geq A_i^{(2)}, \quad i = k+1, k+2, \ldots, n$$

with strict inequality for some index. Then $h^{(1)}(p)$ crosses $h^{(2)}(p)$ at most once for $0 < p < 1$ and if $h^{(1)}(p)$ does cross $h^{(2)}(p)$ for $0 < p < 1$, it does so from below.

Proof. Write

$$h^{(1)}(p) - h^{(2)}(p) = \sum_{i=0}^{n} (A_i^{(1)} - A_i^{(2)}) p^i q^{n-i},$$

with

$$A_i^{(1)} - A_i^{(2)} \leq 0, \quad i = 0, 1, \ldots, k$$
$$\geq 0, \quad i = k+1, k+2, \ldots, n,$$

with strict inequality for at least one value. Let

$$h^{(1)}(p) - h^{(2)}(p) = q^n g(x)$$

where $x = p/q$. Note that g is a polynomial in x; as p traverses the interval $(0, 1)$, x ranges over $(0, \infty)$. By Descartes' rule of signs $g(x)$ has at most one positive real root since the coefficients have at most one sign change. Thus $h^{(1)}(p) - h^{(2)}(p)$ has at most one positive real root for $0 < p < 1$.

Suppose $h^{(1)}(p) - h^{(2)}(p)$ does have a positive real root in $(0, 1)$; then for some i ($0 \leq i \leq k$) $A_i^{(1)} < A_i^{(2)}$. Hence for positive p sufficiently close to 0, $h^{(1)}(p) - h^{(2)}(p) < 0$. Thus $h^{(1)}(p) - h^{(2)}(p)$ is negative to the left of the root and positive to the right. ∥

Lemma 5.1 gives a simple sufficient condition for determining in certain cases which of two structures of a given order yields a "sharper" reliability function. Note that we make no assumption that the structures are monotonic. We use Lemma 5.1 to show that k-out-of-n structures have the sharpest reliability functions among structures of order n. This result generalizes Theorem 2 of E. F. Moore and C. E. Shannon (1956). (Although their quorum function differs slightly from the k-out-of-n function, the proof of the Moore-Shannon Theorem 2 may be carried out in a similar fashion as in the following Theorem 5.2.)

MULTICOMPONENT STRUCTURES

THEOREM 5.2. Let $h(p)$ represent the reliability function of an arbitrary structure of order n and not identical with $h(p; k, n)$, the reliability function of a k-out-of-n structure. Then $h(p)$ and $h(p; k, n)$ intersect in $(0, 1)$ at most once. If they do intersect, $h(p; k, n)$ crosses $h(p)$ from below.

Proof. Write

$$h(p; k, n) = \sum_{i=0}^{n} A_i^{(1)} p^i q^{n-i},$$

where

$$A_i^{(1)} = \begin{cases} 0, & i = 0, 1, \ldots, k-1 \\ \binom{n}{i}, & i = k, k+1, \ldots, n \end{cases}$$

and

$$h(p) = \sum_{i=0}^{n} A_i^{(2)} p^i q^{n-i}.$$

Clearly $A_i^{(1)} \leq A_i^{(2)}$ for $i = 0, 1, \ldots, k-1$ and $A_i^{(1)} \geq A_i^{(2)}$ for $i = k, k+1, \ldots, n$, with strict inequality for some i. The conclusion readily follows from Lemma 5.1. ‖

Finally, if we assume a k-out-of-n structure in which each component life is independently distributed according to F, an IFR distribution with mean μ, we may obtain the following bounds on structure reliability:

$$h(\bar{F}(t)) \begin{cases} \geq \dfrac{n!}{(k-1)!\,(n-k)!} \displaystyle\int_0^{e^{-t/\mu}} x^{k-1}(1-x)^{n-k}\,dx, & t \leq \mu \\[2ex] \leq \dfrac{n!}{(k-1)!\,(n-k)!} \displaystyle\int_0^{1-x_0} x^{k-1}(1-x)^{n-k}\,dx, & t \geq \mu, \end{cases}$$

where x_0 satisfies

$$1 - x_0 = e^{-x_0 t/\mu}$$

(see Theorem 4.5 of Chapter 2). If the first two moments of F are known, the bounds just given may be improved upon by the use of Tables 2 and 3 of Appendix 3.

6. RELATIONSHIP BETWEEN STRUCTURE FAILURE RATE AND COMPONENT FAILURE RATES

Introduction

As we have seen repeatedly, a number of interesting and important consequences follow once we know that a structure has an increasing failure rate. Clearly it would be valuable to have simple sufficient conditions under which a structure has an increasing failure rate when each of its components has an increasing failure rate. In this section we obtain such a condition for structures of like components; for structures composed of components having differing reliabilities, we obtain upper and lower bounds which are increasing functions. In addition, we obtain more general relationships between structure failure rate and component failure rates. The material is based on Esary and Proschan (1963b).

Basic theorem for structures of identical components

We assume that the structure consists of independent like components, with each component life distributed according to the common probability distribution $F(t)$. At a given instant of time t each component has reliability $p = \bar{F}(t)$; the corresponding system reliability will be designated as before by $h(p)$.

The main result of this section is

THEOREM 6.1. Assume a structure with reliability function $h(p)$, with each component life independently distributed according to distribution F having density f. Then

(i) $$\frac{R(t)}{r(t)} = \frac{ph'(p)}{h(p)}\bigg|_{p=\bar{F}(t)},$$

where $r(t) = f(t)/\bar{F}(t)$ denotes component failure rate at time t, and $R(t)$ denotes system failure rate at time t;

(ii) $R(t)/r(t)$ is an increasing function of t if and only if $ph'(p)/h(p)$ is a decreasing function of p;

(iii) if $r(t)$ is an increasing function of t and $ph'(p)/h(p)$ is a decreasing function of p, then $R(t)$ is an increasing function of t.

Result (iii) gives a simple sufficient condition on a structure which will preserve a monotone failure rate when a structure is constructed out of independent like components. We shall present an important class of structures which satisfy this sufficient condition.

To prove (i), let $S(t)$ represent the probability of structure survival past time t; that is, $S(t) = h(\bar{F}(t))$. By definition

$$R(t) = \frac{-S'(t)}{S(t)} = \frac{h'(p)}{h(p)}\bigg|_{p=\bar{F}(t)} \cdot f(t) = \frac{ph'(p)}{h(p)}\bigg|_{p=\bar{F}(t)} \cdot \frac{f(t)}{\bar{F}(t)},$$

so that

$$\frac{R(t)}{r(t)} = \frac{ph'(p)}{h(p)}\bigg|_{p=\bar{F}(t)},$$

establishing (i).

To prove (ii), simply note that $p = \bar{F}(t)$ is a decreasing function of t.

Finally, (iii) is an immediate consequence of (ii) and the fact that p is a decreasing function of t. ∥

An important class of structures for which the condition $ph'(p)/h(p)$ is a decreasing function of p are the k-out-of-n structures discussed in Section 5. To prove that a k-out-of-n structure consisting of n independent components has a ratio $ph'(p)/h(p)$ decreasing in p, write

$$\frac{h(p)}{ph'(p)} = \frac{1}{p} \int_0^p \left(\frac{t}{p}\right)^{k-1} \left(\frac{1-t}{1-p}\right)^{n-k} dt,$$

by (5.2). Letting $u = t/p$, we have

$$\frac{h(p)}{ph'(p)} = \int_0^1 u^{k-1} \left(\frac{1-up}{1-p}\right)^{n-k} du.$$

Since $(1 - up)/(1 - p)$ is increasing in p, so is $h(p)/ph'(p)$. Thus if a k-out-of-n structure is composed of independent like components having an increasing failure rate, the structure itself has an increasing failure rate.

If we note that the time of failure of a k-out-of-n system corresponds to the kth largest in a sample of n observations, an alternate statement of this result is the following.

COROLLARY. *Suppose $X_1 < X_2 < \cdots < X_n$ are order statistics based on independent observations from a distribution having increasing failure rate. Then the distribution of X_i has an increasing failure rate $i = 1, 2, \ldots, n$.*

This result was obtained independently in Chapter 2, Section 4, by a direct calculation. The example next given shows that the result does not necessarily hold if the component distributions are unlike IFR.

Actually, we can generate new structures which have the property that $ph'(p)/h(p)$ is a decreasing function by *composition* of structures having this property. Under composition we form a superstructure each element of which consists of copies of a given structure. If $h = f(g)$ with $g'(p) \geq 0$, then since

$$\frac{ph'(p)}{h(p)} = \frac{gf'(g)}{f(g)} \cdot \frac{pg'(p)}{g(p)},$$

the property is closed under composition.

Applications

For structures composed of independent identical components each having an exponential failure distribution, the application of Theorem 6.1 is particularly simple. Given a structure of like components with common failure distribution $F(t) = 1 - e^{-\lambda t}$, the failure rate $R(t)$ of the structure has as many changes of direction as a function of t as does $ph'(p)/h(p)$ as a function of p; moreover the changes occur in the opposite order. In particular, if $ph'(p)/h(p)$ is decreasing, $R(t)$ is increasing. This follows, since from $r(t) = \lambda$, we have, using Theorem 6.1 (i),

$$R(t) = \lambda \left. \frac{ph'(p)}{h(p)} \right|_{p=\bar{F}(t)}$$

Since $p = \bar{F}(t)$ is a decreasing function of t, the conclusions follow.

Example. From Theorem 6.1 we see that structures composed of like components having exponential failure distributions need not have monotonic failure rates. For example, consider the structure in Figure 6.2

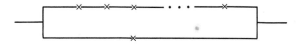

Fig. 6.2

composed of two substructures in parallel, the first having k components in series, the second consisting of a single component. Assuming independent components each having exponential distribution for failure

$$F(t) = 1 - e^{-t},$$

we compute the probability $S(t)$ of structure survival past time t to be

$$S(t) = 1 - (1 - e^{-t})(1 - e^{-kt}).$$

Thus
$$R(t) = -\frac{S'(t)}{S(t)} = \frac{e^{-t} + ke^{-kt} - (k+1)e^{-(k+1)t}}{e^{-t} + e^{-kt} - e^{-(k+1)t}},$$

and so
$$\operatorname{sgn} R'(t) = \operatorname{sgn} [-(k-1)^2 + k^2 e^{-t} + e^{-kt}].$$

Note that given any integer $k > 1$, for $t = 0$ the sign is positive, whereas for $t = \infty$ the sign is negative. Thus the structure failure rate $R(t)$ is *not* monotonic for $k > 1$.

This is also a counterexample to the conjecture that k-out-of-n structures with unlike IFR components are themselves necessarily IFR. (Simply consider the series substructure as a single component.) Another example of a structure having a non-IFR life distribution, even though composed of components having IFR life distributions, is the following. Consider a parallel structure with components i having life distribution F_i ($i = 1, 2, \ldots, n$). Let F_1 be discontinuous at t_0, the end point of its interval of support. Let F_2 be continuous at t_0 with $F_2(t_0) < 1$. Then the system life distribution $\prod_{i=1}^{n} F_i(t)$ is discontinuous at t_0 with $\prod_{i=1}^{n} F_i(t_0) < 1$, so that system life cannot be IFR since an IFR distribution can have a discontinuity only at the right-hand end point of its interval of support.

It is easy to verify that for a structure function having k independent components in series $R(t)/r(t) \equiv k$. Simply note that

$$R(t) = \frac{k[\bar{F}(t)]^{k-1}f(t)}{[\bar{F}(t)]^k} = kr(t).$$

It is interesting that the following converse to this result exists. Assume a structure for which $h(1) = 1$ and $R(t)/r(t) \equiv k$, a constant for $0 \leq t < \infty$. Then it follows that k must be an integer and the structure must consist of k components in series, with the remaining components, if any, nonessential.† To verify this statement, note that $R(t)/r(t) \equiv k$ implies $ph'(p)/h(p) \equiv k$, by Theorem 6.1. Integrating, we have $h(p) = dp^k$, d a constant. Using the assumption $h(1) = 1$, we conclude d must be 1. Thus $h(p) = p^k$. Since the structure consists of a finite number of independent components, $h(p)$ must be a polynomial, and thus k must be an integer. The only finite structure of independent components having reliability function p^k is a series system of k components, with any additional components nonessential.

Bounds on structure failure rate for structures of nonidentical components

Now let us consider a structure of n independent components in which the ith component has probability $p_i(t) = 1 - F_i(t)$ of still being operative at time t ($i = 1, 2, \ldots, n$). If $h(\mathbf{p})$ is the structure reliability function where $\mathbf{p} = (p_1, \ldots, p_n)$, $S(t)$ is the probability of structure survival past time t, and $R(t)$ is the structure failure rate, then since

$$S(t) = h(\mathbf{p}(t)),$$

we obtain

$$-\frac{dS}{dt} = \sum_{i=1}^n \frac{\partial h}{\partial p_i}\left(-\frac{dp_i}{dt}\right) = \sum_{i=1}^n \frac{\partial h}{\partial p_i} \cdot f_i(t).$$

Hence

$$R(t) = \frac{-dS/dt}{S(t)} = \sum_{i=1}^n \frac{\bar{F}_i(t)}{S(t)} \cdot \frac{\partial h}{\partial p_i} \cdot \frac{f_i(t)}{\bar{F}_i(t)} = \sum_{i=1}^n \frac{p_i \, \partial h/\partial p_i}{h(\mathbf{p})} r_i(t). \quad (6.1)$$

Equation (6.1) states that the structure failure rate is the inner product of two vectors: the vector whose ith component is

$$\frac{p_i \, \partial h/\partial p_i}{h(\mathbf{p})}$$

and the vector whose ith component is $r_i(t)$. Note that the first vector is a function of the structure regardless of component failure rates, whereas

† A nonessential component is one that may be omitted with no effect on structure performance.

the second vector is a function of component failure rates regardless of structure.

Using (6.1) we may obtain bounds on structure failure rate $R(t)$ as follows:

$$\left[\sum_{i=1}^{n} \frac{p_i \,\partial h/\partial p_i}{h(\mathbf{p})}\right] \min_{1\leq i \leq n} r_i(t) \leq R(t) \leq \left[\sum_{i=1}^{n} \frac{p_i \,\partial h/\partial p_i}{h(\mathbf{p})}\right] \max_{1\leq i \leq n} r_i(t). \quad (6.2)$$

Next we shall show that for k-out-of-n structures the factor $\sum_{i=1}^{n} \frac{p_i \partial h/\partial p_i}{h(\mathbf{p})}$ is strictly decreasing in each of $p_1, \ldots, p_n$. Thus for k-out-of-n structures we may bracket the structure failure rate $R(t)$ from below and above by a pair of bounds which are strictly increasing if component failure rates are increasing. This result is especially interesting since, by the example of this section, we know that $R(t)$ is not necessarily increasing for k-out-of-n structures with components having *different* IFR distributions.

THEOREM 6.2. Let $h(\mathbf{p})$ be the reliability function of a k-out-of-n structure of independent components. Then

$$u(\mathbf{p}) = \sum_{i=1}^{n} \frac{p_i \,\partial h/\partial p_i}{h(\mathbf{p})}$$

is strictly decreasing in $0 \leq p_i \leq 1$ $(i = 1, 2, \ldots, n)$.

Proof. We may expand $h(\mathbf{p})$ as

$$h(\mathbf{p}) = p_i h(1_i, \mathbf{p}) + (1 - p_i) h(0_i, \mathbf{p}),$$

where $h(1_i, \mathbf{p})$ is the conditional reliability given the ith component is operating, and $h(0_i, \mathbf{p})$ is the conditional reliability given the ith component has failed. It follows that

$$\frac{\partial h}{\partial p_i} = h(1_i, \mathbf{p}) - h(0_i, \mathbf{p}) = P\left[\sum_{j \neq i} X_j \geq k - 1\right] - P\left[\sum_{j \neq i} X_j \geq k\right]$$

where $X_j = 1$ or 0 according as the jth component is operating or not. Hence

$$\frac{\partial h}{\partial p_i} = P\left[\sum_{j \neq i} X_j = k - 1\right].$$

Thus $\sum_{i=1}^{n} p_i \frac{\partial h}{\partial p_i} = \sum_{i=1}^{n} P\left[X_i = 1, \sum_{j \neq i} X_j = k - 1\right] = kP[\sum X_j = k]$

since each term used in computing $\Sigma X_j = k$ will occur exactly k times in computing $\sum_{i=1}^{n} P\left[X_i = 1, \sum_{j \neq i} X_j = k - 1\right]$.

It will thus be sufficient to prove

$$\frac{P[\sum X_j = k]}{P[\sum X_j \geq k]}$$

is strictly decreasing in p_i ($i = 1, 2, \ldots, n$) or equivalently, that

$$\frac{P[\sum X_j \geq k + 1]}{P[\sum X_j \geq k]}$$

is strictly increasing in p_i ($i = 1, 2, \ldots, n$). This is equivalent to proving that for $p_i < p_i'$

$$0 > \begin{vmatrix} P[\sum X_j \geq k+1 \,|\, p_i] & P[\sum X_j \geq k+1 \,|\, p_i'] \\ P[\sum X_j \geq k \,|\, p_i] & P[\sum X_j \geq k \,|\, p_i'] \end{vmatrix}$$

$$= \begin{vmatrix} p_i P\left[\sum_{j \neq i} X_j \geq k\right] + q_i P\left[\sum_{j \neq i} X_j \geq k+1\right] & p_i' P\left[\sum_{j \neq i} X_j \geq k\right] + q_i' P\left[\sum_{j \neq i} X_j \geq k+1\right] \\ p_i P\left[\sum_{j \neq i} X_j \geq k-1\right] + q_i P\left[\sum_{j \neq i} X_j \geq k\right] & p_i' P\left[\sum_{j \neq i} X_j \geq k-1\right] + q_i' P\left[\sum_{j \neq i} X_j \geq k\right] \end{vmatrix}$$

$$= (p_i q_i' - p_i' q_i) \begin{vmatrix} P\left[\sum_{j \neq i} X_j \geq k\right] & P\left[\sum_{j \neq i} X_j \geq k+1\right] \\ P\left[\sum_{j \neq i} X_j \geq k-1\right] & P\left[\sum_{j \neq i} X_j \geq k\right] \end{vmatrix}$$

But $p_i q_i' - p_i' q_i < 0$. Moreover, it is easy to verify that $P\left[\sum_{j \neq i} X_j \geq k\right]$ is totally positive of order infinity in differences of k (Karlin and Proschan, 1960), so that the determinant just given is nonnegative; actually it can be proved positive. Thus the desired conclusion follows. ‖

APPENDIX 1
Total Positivity

In this appendix we summarize some of the properties of totally positive functions helpful in understanding the implications of such concepts as increasing failure rate, decreasing failure rate, etc. Many of the most useful consequences of these concepts derive from the happy fact that distributions with monotone failure rate are members of the class of totally positive functions.

The theory of totally positive functions has been extensively applied in mathematics, statistics, economics, and mechanics. Totally positive kernels arise naturally in developing procedures for inverting by differential polynomial operators the integral transformations defined in terms of convolution kernels (Hirschman and Widder, 1955). The theory of total positivity is fundamental in permitting characterizations of best statistical procedures for decision problems (Karlin, 1956, Karlin, 1957, and Karlin and Rubin, 1956). In clarifying the structure of stochastic processes with continuous-path functions, we encounter totally positive kernels (Karlin and McGregor, 1959a,b). Studies in the stability of certain models in mathematical economics frequently use properties of totally positive kernels (Karlin, 1959, and Arrow, Karlin, and Scarf, 1958). The theory of vibrations of certain types of mechanical systems involves aspects of the theory of totally positive kernels (Gantmacher and Krein, 1950). Finally, as we hope is demonstrated throughout this monograph, total positivity of order 2 (specifically, monotone failure rate) is a very fruitful concept in reliability theory.

A function $f(x, y)$ of two variables ranging over linearly ordered one-dimensional sets X and Y, respectively, is said to be *totally positive of order* k (TP_k) if for all $x_1 < x_2 < \cdots < x_m$, $y_1 < y_2 < \cdots < y_m$ ($x_i \in X$, $y_j \in Y$), and all $1 \leq m \leq k$,

$$f\begin{pmatrix} x_1, x_2, \ldots, x_m \\ y_1, y_2, \ldots, y_m \end{pmatrix} \equiv \begin{vmatrix} f(x_1, y_1) & f(x_1, y_2) & \cdots & f(x_1, y_m) \\ f(x_2, y_1) & f(x_2, y_2) & \cdots & f(x_2, y_m) \\ \cdot & \cdot & \cdot & \cdot \\ \cdot & \cdot & \cdot & \cdot \\ \cdot & \cdot & \cdot & \cdot \\ f(x_m, y_1) & f(x_m, y_2) & \cdots & f(x_m, y_m) \end{vmatrix} \geq 0. \quad (1)$$

Typically, X is an interval of the real line or a countable set of discrete values on the real line such as the set of all integers or the set of nonnegative integers; this is also true for Y. When X or Y is a set of integers, we may use the term "sequence" rather than "function."

A related concept is that of sign reverse regularity. A function $f(x, y)$ is *reverse regular of order* k (RR_k) if for every $x_1 < x_2 < \cdots < x_m$, $y_1 < y_2 < \cdots < y_m$, and $1 \leq m \leq k$,

$$(-1)^{m(m-1)/2} f\begin{pmatrix} x_1, x_2, \ldots, x_m \\ y_1, y_2, \ldots, y_m \end{pmatrix} \geq 0.$$

If a TP_k function is a probability density in one of the variables, say x, with respect to a σ-finite measure $\mu(x)$ for each fixed value of y, and is expressible as a function $f(x - y)$ of the difference of x and y, then f is said to be a *Pólya frequency function* (or density) *of order* k (PF_k). The argument traverses the real line. If the argument is confined to the integers, we shall speak of a *Pólya frequency sequence of order* k (PF_k sequence).

When the subscript ∞ is attached to any of the definitions, the property in question will be understood to hold for all positive integers. The term "strict" is used to mean that the inequalities in question are strict.

Many of the structural properties of totally positive functions are deducible from the following identity which appears in Pólya and Szegö (1925, p. 48 prob. 68).

LEMMA 1. *If* $r(x, w) = \int p(x, t) q(t, w) \, d\sigma(t)$ *and the integral converges absolutely, then*

$$r\begin{pmatrix} x_1, x_2, \ldots, x_k \\ w_1, w_2, \ldots, w_k \end{pmatrix}$$
$$= \iint \cdots \int_{t_1 < t_2 < \cdots < t_k} p\begin{pmatrix} x_1, x_2, \ldots, x_k \\ t_1, t_2, \ldots, t_k \end{pmatrix} q\begin{pmatrix} t_1, t_2, \ldots, t_k \\ w_1, w_2, \ldots, w_k \end{pmatrix} d\sigma(t_1) d\sigma(t_2) \cdots d\sigma(t_k).$$

For the case in which the arguments occur as differences, a proof may be found in Schoenberg (1951, pp. 341–343) and in Hirschman and Widder (1955). In particular, we obtain from Lemma 1 the following useful result.

LEMMA 2. *If* $f(x, t)$ *is* TP_m *and* $g(t, w)$ *is* TP_n, *then*

$$h(x, w) = \int f(x, t) g(t, w) \, d\sigma(t)$$

is $TP_{\min(m, n)}$ *provided* $\sigma(t)$ *is a regular* σ-*finite measure.*

An important feature of totally positive functions is their variation diminishing property.

APPENDIX 1

VARIATION DIMINISHING PROPERTY

Suppose $K(x, y)$ is TP_r and $f(y)$ changes sign† $j \leq r - 1$ times. Let $g(x) = \int K(x, y) f(y) \, d\mu(y)$, an absolutely convergent integral with μ a σ-finite measure. Then $g(x)$ changes sign at most j times. Moreover, if $g(x)$ actually changes sign j times, then $g(x)$ must have the same arrangement of signs as does $f(y)$ as x and y traverse their respective domains from left to right.

The variation diminishing property is actually equivalent to the inequalities (1). This property underlies many of the applications of totally positive functions. See Schoenberg (1951 and 1953) for an exhaustive study of the variation diminishing properties of totally positive kernels, where the variables x and y occur in translation form. See Karlin (1963) for a study and applications of the case where the variables x and y are separate.

Many of the properties of the totally positive functions of specific interest in reliability theory have been developed as needed throughout the monograph, for example, the properties of distributions with monotone failure rate. In the following we develop some additional properties of PF_2 functions especially needed in the checking model of Section 3, Chapter 4.

First we prove the equivalence of the two definitions of PF_2 functions given in Section 4, Chapter 2.

THEOREM 1. A density f is PF_2 (as just defined) if and only if for all Δ,

$$\frac{F(x + \Delta) - F(x)}{f(x)} \text{ is decreasing in } x. \tag{2}$$

Proof. (2) is equivalent to the condition that for all $\Delta \geq 0$

$$\frac{F(x + \Delta) - F(x)}{f(x)} \text{ is decreasing in } x, \tag{3}$$

and for all $\Delta \geq 0$

$$\frac{F(x) - F(x - \Delta)}{f(x)} \text{ is increasing in } x. \tag{4}$$

First we show that (3) and (4) imply f is PF_2.

† The number of sign changes $V(f)$ of a real valued function f is sup $V[f(x_i)]$, where $V[f(x_i)]$ denotes the number of sign changes of the sequence $f(x_1), f(x_2), \ldots, f(x_m)$ (zero values are discarded), with x_i chosen arbitrarily from the domain of definition of f and arranged so that $x_1 < x_2 < \cdots < x_m$, m any positive integer.

Statement (4) implies for $x < y$

$$\frac{f(x-\Delta)}{f(x)}\left[\frac{F(x)-F(x-\Delta)}{f(x-\Delta)}\right] \leq \frac{f(y-\Delta)}{f(y)}\left[\frac{F(y)-F(y-\Delta)}{f(y-\Delta)}\right].$$

Using (3) this implies

$$\frac{f(x-\Delta)}{f(x)} \leq \frac{f(y-\Delta)}{f(y)}$$

which completes the argument.

If f is a PF$_2$ density, then for $x < y$

$$\frac{f(x+t)}{f(x)} \geq \frac{f(y+t)}{f(y)}$$

for $t > 0$. This implies

$$\int_0^\Delta \frac{f(x+t)}{f(x)}\,dt \geq \int_0^\Delta \frac{f(y+t)}{f(y)}\,dt,$$

so that

$$\frac{F(x+\Delta)-F(x)}{f(x)} \geq \frac{F(y+\Delta)-F(y)}{f(y)}.$$

Moreover, for $x < y$

$$\frac{f(x-t)}{f(x)} \leq \frac{f(y-t)}{f(y)}$$

which implies

$$\int_0^\Delta \frac{f(x-t)}{f(x)}\,dt \leq \int_0^\Delta \frac{f(y-t)}{f(y)}\,dt$$

yielding (4). ‖

An important shape property of PF$_2$ densities is their unimodality. We shall find this useful in Theorem 4. The proof is based on Karlin (in preparation).

THEOREM 2. If $f(u)$ is PF$_2$, then $f(u)$ is unimodal.

To avoid technical details, we assume that $f(u) > 0$ and $f'(u)$ exists everywhere. By subtracting the first column from the second in the determinantal inequality for PF$_2$ functions and taking limits, we have

$$\begin{vmatrix} f(x_1-y) & -f'(x_1-y) \\ f(x_2-y) & -f'(x_2-y) \end{vmatrix} \geq 0$$

for $x_1 < x_2$ and y arbitrary. If for some u_0, $f'(u_0) = 0$, we deduce $f'(u) \geq 0$ for $u < u_0$ and $f'(u) \leq 0$ for $u > u_0$. Thus $f(u)$ is unimodal. ‖

In the checking model of Section 3, Chapter 4 we need the following result.

THEOREM 3. If f is PF_2, $x < y$, $r \geq 1$, $\Delta \geq 0$, then

$$\frac{F(y) - F(y - r\Delta)}{f(y)} \geq r \frac{F(x) - F(x - \Delta)}{f(x)}, \quad x - r\Delta \geq m,$$

$$\frac{F(x + r\Delta) - F(x)}{f(x)} \geq r \frac{F(y + \Delta) - F(y)}{f(y)}, \quad y + r\Delta \leq m,$$

where m is the mode of f.

Proof. We will prove only the first inequality since the proof is similar for the second. Since f is unimodal by Theorem 2, $y - r\Delta \geq m$, and $r \geq 1$, it follows that

$$\frac{\int_0^{r\Delta} f(y - t) \, dt}{\int_0^{\Delta} f(y - t) \, dt} \geq r.$$

For $x < y$,

$$\int_0^{r\Delta} \frac{f(y - t)}{f(y)} \, dt \geq r \int_0^{\Delta} \frac{f(y - t)}{f(y)} \, dt \geq r \int_0^{\Delta} \frac{f(x - t)}{f(x)} \, dt$$

since $f(x - t)/f(x)$ is increasing in x for $t > 0$, as observed in (4.4) of Chapter 2. ∥

APPENDIX 2
Test for Increasing Failure Rate

We have seen the important role that distributions with increasing failure rate play in reliability theory. Therefore it would be of great value to have a test to determine whether a sample (or set of samples) comes from a population with IFR. In this appendix we present such a test (Proschan and Pyke, in preparation), although this treatment of a statistical problem represents an exception from the restriction followed throughout the book of treating only probabilistic problems, that is, problems involving known distributions.

Let $X_1, X_2, \ldots, X_n$ be a sample of independent observations from the common distribution F with density f, where $f(t) = 0$ for $t < 0$, and failure rate $r(t)$. We wish to choose between the following:

Null hypothesis, H_0: r is constant.
Alternative hypothesis, H_1: r is nondecreasing but not constant.

The test statistic is computed as follows. Let $T_1 < T_2 < \cdots < T_n$ be the ordered observations, $D_1 = T_1, D_2 = T_2 - T_1, \ldots, D_n = T_n - T_{n-1}$ the spacings, and $\bar{D}_1 = nD_1, \bar{D}_2 = (n-1)D_2, \ldots, \bar{D}_n = D_n$ the normalized spacings. For $i, j = 1, 2, \ldots, n$ let $V_{ij} = 1$ if $\bar{D}_i > \bar{D}_j$, 0 otherwise. The test statistic is

$$V_n = \sum_{\substack{i,j=1 \\ i<j}}^{n} V_{ij}. \tag{1}$$

We reject the null hypothesis at the α level of significance if $V_n > v_{n,\alpha}$ where $v_{n,\alpha}$ is determined such that $P[V_n > v_{n,\alpha} \mid H_0] = \alpha$.

Heuristically we may justify the test as follows. Under the null hypothesis it may be readily verified that $\bar{D}_1, \bar{D}_2, \ldots, \bar{D}_n$ are independent exponential random variables with common parameter, say λ. Hence $P[V_{ij} = 1] = \frac{1}{2}$, for $i, j = 1, 2, \ldots, n, i \neq j$. However, as will be shown, under the alternative hypothesis, $P[V_{ij} = 1] > \frac{1}{2}$ for $i < j, i, j = 1, 2, \ldots, n$. In fact, each V_{ij} and V_n tends to be larger under the alternative hypothesis, so that rejection of the null hypothesis occurs for large values of V_n. Since under the null hypothesis the distribution of V_n is known, we have available $v_{n,\alpha}$.

DISTRIBUTION UNDER THE NULL HYPOTHESIS

Since under H_0, $\bar{D}_1, \bar{D}_2, \ldots, \bar{D}_n$ are independently distributed, each having density $\lambda e^{-\lambda t}$, all orderings of $\bar{D}_1, \bar{D}_2, \ldots, \bar{D}_n$ are equally likely. Using this property, Kendall (1938) provides tables for $P[V_n \leq k \mid H_0]$, $n \leq 10$; more convenient tables are available in Mann (1945). Mann shows that μ_n and σ_n^2, the mean and variance of V_n, are given by

$$\mu_n = \frac{n(n-1)}{4}, \qquad \sigma_n^2 = \frac{(2n+5)(n-1)n}{72}, \qquad (2)$$

and that V_n is asymptotically normal.

UNBIASEDNESS OF TEST

We now show that V_n is unbiased, that is, that $P[V_n \geq v_{n,\alpha} \mid H_1] \geq \alpha$ for $0 < \alpha \leq 1$, $n = 2, 3, \ldots$. Make the transformation

$$X_i' = -\log \bar{F}(X_i). \qquad (3)$$

It follows that

$$P[X_i' > u] = e^{-u}.$$

Thus each X_i' is distributed according to the exponential distribution with unit mean. Moreover, since the $X_1, \ldots, X_n$ are independent, so are the $X_1', \ldots, X_n'$. Next let $T_1' < T_2' < \cdots < T_n'$ represent the ranked $X_1', X_2', \ldots, X_n'$ so that $T_i' = -\log \bar{F}(T_i)$ for $i = 1, 2, \ldots, n$. Further, let

$$\bar{D}_i' = (n - i + 1)(T_i' - T_{i-1}'), \qquad i = 1, 2, \ldots, n,$$

where $T_0' = 0$ by definition. It is easy to verify that the $\bar{D}_i'$ are independently, identically distributed according to the exponential with unit mean (see, for example, Epstein and Sobel, 1953 or Rényi, 1953).

Note that T_i' is an increasing function of T_i. Moreover T_i' is a convex function of T_i as shown in Theorem 4.1 of Chapter 2. It follows that $\bar{D}_i' \geq \bar{D}_j'$ implies $\bar{D}_i \geq \bar{D}_j$ for $i < j$. Thus $V_{ij} \geq V_{ij}'$ where $V_{ij}' = 1$ if $\bar{D}_i' \geq \bar{D}_j'$. Hence $V_n \geq V_n'$, where $V_n' = \sum_{i<j} V_{ij}'$, so that

$$P[V_n \geq v_{n,\alpha} \mid H_1] \geq \alpha$$

for $0 < \alpha \leq 1$, $n = 2, 3, \ldots$, implying that V_n is unbiased, as claimed.

ASYMPTOTIC DISTRIBUTION UNDER THE ALTERNATIVE HYPOTHESIS

Under the alternative hypothesis, the asymptotic distribution of V_n is normal under mild restrictions (Proschan and Pyke, in preparation). It follows that V_n is a consistent test.

ASYMPTOTIC RELATIVE EFFICIENCY

To compare the V_n test against various possible competing tests, we compute its asymptotic relative efficiency (ARE). The ARE of V_n against a competing test statistic T_n which has a normal distribution asymptotically with variance of order n^{-1} may be expressed, under certain regularity conditions, as

$$\frac{\mu'(\theta_o)/\sigma^2(\theta_o)}{\mu_T'(\theta_o)/\sigma^2(\theta_o)}, \tag{4}$$

where $\mu_T(\theta)$, $\sigma_T^2(\theta)$ are the limiting mean and variance, respectively, of the test statistic T_n corresponding to a parameter value θ, and $\mu(\theta)$, $\sigma^2(\theta)$ are the limiting mean and variance of V_n; θ_o represents the parameter value for which the alternative hypothesis coincides with the null hypothesis. Roughly speaking, the asymptotic relative efficiency of V_n versus T_n measures the relative number of observations required using T_n to achieve the variance possessed by V_n asymptotically.

(a) *Likelihood ratio test against the Weibull distribution.* Let the distribution under H_1 be the Weibull with IFR:

$$F(x) = 1 - e^{-\lambda x^\theta}, \quad \lambda > 0, \theta > 1, x \geq 0, \tag{5}$$

with λ assumed known. The likelihood ratio test can be shown after simplification to be

$$T_n^w = \sum_{i=1}^n (1 - \lambda x_i) \log X_i,$$

with

$$\text{ARE of } V_n \text{ versus } T_n^w = \frac{1.0809}{(\log \lambda - .4228)^2 + 1.6449}$$

which achieves a maximum value of .6571 and $\to 0$ as $\lambda \to 0$ or ∞.

(b) *Likelihood ratio test against the gamma distribution.* Let the distribution under H_1 be the gamma distribution with IFR, with density

$$f(x) = \frac{\lambda^\theta x^{\theta-1} e^{-\lambda x}}{\Gamma(\theta)}, \quad \theta \geq 1. \tag{6}$$

The likelihood ratio test statistic is

$$T_n^G = \sum_{i=1}^{n} \log X_i$$

with   ARE of V_n versus $T_n^G = .2040$

independent of λ.

(c) *Coefficient of variation test.* We have seen in Section 4 of Chapter 2 that the coefficient of variation $c = \sigma/\mu \leq 1$ for F (IFR), whereas $c = 1$ for F exponential. As pointed out by S. Karlin, this suggests the test statistic $C(\bar{X}, s) = s/\bar{X}$, where $\bar{X} = \frac{1}{n}\sum_{i=1}^{n} X_i$ and $s^2 = \frac{1}{n}\sum_{i=1}^{n}(X_i - \bar{X})^2$ are the sample mean and variance. Proschan and Pyke (in preparation) show

ARE of V_n versus $C = 1.0809$

when the underlying distribution is Weibull (5) and

ARE of V_n versus $C = 4.3236$

when the underlying distribution is gamma (6).

Other test statistics are considered in Proschan and Pyke (in preparation).

APPENDIX 3

Tables Giving Bounds on Distributions with Monotone Failure Rate

TABLE 1

UPPER BOUNDS ON $1 - F(t)$

$$\left(F \text{ is IFR}, \mu_1 = \int_0^\infty t \, dF(t) = 1 \right)$$

t	IFR Bound	Markov Bound $(1/t)$	t	IFR Bound	Markov Bound $(1/t)$
1.0	1.000	1.000			
1.1	0.824	0.909	3.1	0.053	0.323
1.2	0.686	0.833	3.2	0.047	0.313
1.3	0.577	0.769	3.3	0.042	0.303
1.4	0.489	0.714	3.4	0.038	0.294
1.5	0.417	0.667	3.5	0.034	0.286
1.6	0.358	0.625	3.6	0.030	0.278
1.7	0.309	0.588	3.7	0.027	0.270
1.8	0.268	0.555	3.8	0.025	0.263
1.9	0.233	0.526	3.9	0.022	0.256
2.0	0.203	0.500	4.0	0.020	0.250
2.1	0.178	0.476	4.1	0.018	0.244
2.2	0.156	0.455	4.2	0.016	0.238
2.3	0.138	0.435	4.3	0.015	0.233
2.4	0.121	0.417	4.4	0.013	0.227
2.5	0.107	0.400	4.5	0.012	0.222
2.6	0.095	0.385	4.6	0.011	0.217
2.7	0.084	0.370	4.7	0.010	0.213
2.8	0.075	0.357	4.8	0.009	0.208
2.9	0.067	0.345	4.9	0.008	0.204
3.0	0.060	0.333	5.0	0.007	0.200

TABLE 2

UPPER BOUNDS ON $1 - F(t)$

$$\left(F \text{ is IFR}, \mu_1 = \int_0^\infty t\, dF(t) = 1, \mu_2 = \int_0^\infty t^2\, dF(t)\right)$$

	μ_2									
t	1.1	1.2	1.3	1.4	1.5	1.6	1.7	1.8	1.9	2.0
0.1	1.000	1.000	1.000	1.000	1.000	1.000	1.000	1.000	0.953	0.905
0.2	1.000	1.000	1.000	1.000	1.000	1.000	0.965	0.911	0.862	0.819
0.3	1.000	1.000	1.000	1.000	0.994	0.929	0.873	0.824	0.780	0.741
0.4	1.000	1.000	1.000	0.970	0.901	0.842	0.791	0.746	0.706	0.671
0.5	1.000	1.000	0.957	0.881	0.817	0.763	0.716	0.675	0.639	0.607
0.6	1.000	0.959	0.871	0.800	0.741	0.691	0.648	0.611	0.578	0.549
0.7	0.988	0.878	0.794	0.727	0.672	0.626	0.587	0.553	0.523	0.497
0.8	0.914	0.805	0.725	0.661	0.610	0.568	0.532	0.501	0.474	0.450
0.9	0.850	0.740	0.662	0.602	0.554	0.515	0.482	0.453	0.429	0.407
1.0	0.797	0.683	0.605	0.548	0.503	0.467	0.436	0.410	0.388	0.368
1.1	0.756	0.632	0.555	0.499	0.457	0.423	0.395	0.371	0.351	0.333
1.2	0.633	0.589	0.509	0.456	0.415	0.384	0.358	0.336	0.318	0.302
1.3	0.411	0.555	0.469	0.416	0.378	0.348	0.324	0.304	0.288	0.273
1.4	0.259	0.444	0.434	0.381	0.344	0.316	0.294	0.276	0.260	0.247
1.5	0.163	0.321	0.405	0.349	0.313	0.287	0.266	0.250	0.235	0.224
1.6	0.105	0.231	0.332	0.321	0.285	0.260	0.241	0.226	0.213	0.202
1.7	0.068	0.167	0.254	0.296	0.260	0.237	0.219	0.205	0.193	0.183
1.8	0.046	0.122	0.194	0.259	0.238	0.215	0.198	0.185	0.175	0.165
1.9	0.031	0.089	0.149	0.205	0.218	0.196	0.180	0.168	0.158	0.150
2.0	0.021	0.066	0.116	0.163	0.201	0.178	0.163	0.152	0.143	0.136
2.1	0.015	0.050	0.090	0.130	0.169	0.162	0.148	0.138	0.130	0.123
2.2	0.010	0.038	0.070	0.105	0.138	0.148	0.134	0.125	0.117	0.111
2.3	0.007	0.029	0.056	0.084	0.114	0.135	0.122	0.113	0.106	0.101
2.4	0.005	0.022	0.044	0.068	0.094	0.119	0.111	0.102	0.096	0.091
2.5	0.004	0.017	0.035	0.056	0.077	0.099	0.101	0.093	0.087	0.083
2.6	0.003	0.013	0.028	0.046	0.064	0.083	0.092	0.084	0.079	0.075
2.7	0.002	0.010	0.023	0.037	0.053	0.070	0.084	0.076	0.071	0.068
2.8	0.002	0.008	0.018	0.031	0.045	0.059	0.074	0.069	0.065	0.061
2.9	0.001	0.006	0.015	0.026	0.037	0.050	0.063	0.063	0.059	0.055
3.0	0.001	0.005	0.012	0.021	0.031	0.043	0.054	0.057	0.053	0.050

TABLE 3

LOWER BOUNDS ON $1 - F(t)$

$$\left(F \text{ is IFR}, \mu_1 = \int_0^\infty t\, dF(t) = 1, \mu_2 = \int_0^\infty t^2\, dF(t)\right)$$

	μ_2								
t	1.1	1.2	1.3	1.4	1.5	1.6	1.7	1.8	1.9
0.1	0.974	0.955	0.941	0.930	0.922	0.916	0.912	0.908	0.906
0.2	0.949	0.913	0.886	0.866	0.851	0.840	0.831	0.825	0.821
0.3	0.925	0.872	0.834	0.806	0.785	0.770	0.758	0.750	0.744
0.4	0.900	0.833	0.785	0.750	0.724	0.705	0.691	0.681	0.674
0.5	0.868	0.789	0.736	0.698	0.668	0.646	0.630	0.619	0.611
0.6	0.819	0.731	0.677	0.640	0.613	0.592	0.575	0.562	0.553
0.7	0.747	0.655	0.605	0.573	0.551	0.534	0.521	0.510	0.501
0.8	0.640	0.561	0.524	0.501	0.486	0.475	0.466	0.459	0.453
0.9	0.501	0.459	0.441	0.430	0.423	0.418	0.414	0.411	0.408
1.0	0.367	0.367	0.367	0.367	0.367	0.367	0.367	0.367	0.367
1.1	0.269	0.294	0.306	0.314	0.319	0.323	0.326	0.329	0.331
1.2	0.000	0.235	0.255	0.268	0.277	0.284	0.289	0.294	0.298
1.3	0.000	0.188	0.213	0.229	0.240	0.249	0.257	0.263	0.268
1.4	0.000	0.000	0.177	0.195	0.209	0.219	0.228	0.235	0.241
1.5	0.000	0.000	0.145	0.166	0.181	0.193	0.202	0.210	0.217
1.6	0.000	0.000	0.000	0.141	0.157	0.169	0.179	0.188	0.195
1.7	0.000	0.000	0.000	0.117	0.136	0.149	0.159	0.168	0.176
1.8	0.000	0.000	0.000	0.000	0.116	0.130	0.141	0.150	0.158
1.9	0.000	0.000	0.000	0.000	0.098	0.114	0.125	0.134	0.142
2.0	0.000	0.000	0.000	0.000	0.078	0.098	0.110	0.120	0.128
2.1	0.000	0.000	0.000	0.000	0.000	0.084	0.097	0.107	0.115
2.2	0.000	0.000	0.000	0.000	0.000	0.071	0.085	0.095	0.104
2.3	0.000	0.000	0.000	0.000	0.000	0.057	0.074	0.084	0.093
2.4	0.000	0.000	0.000	0.000	0.000	0.000	0.064	0.075	0.084
2.5	0.000	0.000	0.000	0.000	0.000	0.000	0.055	0.066	0.075
2.6	0.000	0.000	0.000	0.000	0.000	0.000	0.046	0.058	0.067
2.7	0.000	0.000	0.000	0.000	0.000	0.000	0.037	0.051	0.060
2.8	0.000	0.000	0.000	0.000	0.000	0.000	0.000	0.045	0.054
2.9	0.000	0.000	0.000	0.000	0.000	0.000	0.000	0.039	0.048
3.0	0.000	0.000	0.000	0.000	0.000	0.000	0.000	0.033	0.043

TABLE 4

LOWER BOUNDS ON $1 - F(t)$

$$\left(F \text{ is DFR}, \mu_1 = \int_0^\infty t\, dF(t) = 1, \mu_2 = \int_0^\infty t^2\, dF(t)\right)$$

μ_2

t	2.1	2.2	2.3	2.4	2.5	2.6	2.7	2.8	2.9	3.0
0.0	0.952	0.909	0.869	0.833	0.800	0.769	0.740	0.714	0.689	0.666
0.1	0.861	0.822	0.787	0.754	0.724	0.696	0.671	0.647	0.625	0.604
0.2	0.779	0.744	0.712	0.683	0.656	0.631	0.608	0.586	0.566	0.548
0.3	0.705	0.673	0.644	0.618	0.594	0.571	0.551	0.531	0.514	0.497
0.4	0.638	0.609	0.583	0.559	0.538	0.517	0.499	0.482	0.466	0.451
0.5	0.577	0.551	0.528	0.506	0.487	0.469	0.452	0.437	0.422	0.409
0.6	0.522	0.499	0.478	0.458	0.441	0.425	0.410	0.396	0.383	0.371
0.7	0.473	0.452	0.432	0.415	0.399	0.385	0.371	0.359	0.347	0.336
0.8	0.428	0.409	0.391	0.376	0.362	0.348	0.336	0.325	0.315	0.305
0.9	0.387	0.370	0.354	0.340	0.327	0.316	0.305	0.295	0.286	0.277
1.0	0.350	0.335	0.321	0.308	0.297	0.286	0.276	0.267	0.259	0.251
1.1	0.317	0.303	0.290	0.279	0.269	0.259	0.250	0.242	0.235	0.228
1.2	0.287	0.274	0.263	0.253	0.243	0.235	0.227	0.220	0.213	0.207
1.3	0.259	0.248	0.238	0.229	0.220	0.213	0.206	0.199	0.193	0.188
1.4	0.235	0.224	0.215	0.207	0.200	0.193	0.186	0.181	0.175	0.170
1.5	0.212	0.203	0.195	0.187	0.181	0.175	0.169	0.164	0.159	0.155
1.6	0.192	0.184	0.176	0.170	0.164	0.158	0.153	0.149	0.144	0.140
1.7	0.174	0.166	0.160	0.154	0.148	0.143	0.139	0.135	0.131	0.127
1.8	0.157	0.150	0.144	0.139	0.134	0.130	0.126	0.122	0.119	0.116
1.9	0.142	0.136	0.131	0.126	0.122	0.118	0.114	0.111	0.108	0.105
2.0	0.129	0.123	0.118	0.114	0.110	0.107	0.104	0.101	0.098	0.095
2.1	0.116	0.111	0.107	0.103	0.100	0.097	0.094	0.091	0.089	0.087
2.2	0.105	0.101	0.097	0.094	0.090	0.088	0.085	0.083	0.081	0.079
2.3	0.095	0.091	0.088	0.085	0.082	0.079	0.077	0.075	0.073	0.071
2.4	0.086	0.083	0.079	0.077	0.074	0.072	0.070	0.068	0.066	0.065
2.5	0.078	0.075	0.072	0.069	0.067	0.065	0.063	0.062	0.060	0.059
2.6	0.070	0.068	0.065	0.063	0.061	0.059	0.057	0.056	0.055	0.053
2.7	0.064	0.061	0.059	0.057	0.055	0.054	0.052	0.051	0.050	0.048
2.8	0.058	0.055	0.053	0.052	0.050	0.049	0.047	0.046	0.045	0.044
2.9	0.052	0.050	0.048	0.047	0.045	0.044	0.043	0.042	0.041	0.040
3.0	0.047	0.045	0.044	0.042	0.041	0.040	0.039	0.038	0.037	0.036

TABLE 4 (continued)
LOWER BOUNDS ON $1 - F(t)$

$$\left(F \text{ is DFR}, \mu_1 = \int_0^\infty t\, dF(t) = 1, \mu_2 = \int_0^\infty t^2\, dF(t)\right)$$

	μ_2									
t	3.1	3.2	3.3	3.4	3.5	3.6	3.7	3.8	3.9	4.0
0.0	0.645	0.625	0.606	0.588	0.571	0.555	0.540	0.526	0.512	0.500
0.1	0.585	0.567	0.550	0.534	0.519	0.504	0.491	0.478	0.466	0.455
0.2	0.531	0.514	0.499	0.485	0.471	0.458	0.446	0.435	0.424	0.414
0.3	0.481	0.467	0.453	0.440	0.428	0.417	0.406	0.396	0.386	0.377
0.4	0.437	0.424	0.411	0.400	0.389	0.379	0.369	0.360	0.351	0.343
0.5	0.396	0.385	0.374	0.363	0.354	0.344	0.336	0.327	0.320	0.312
0.6	0.360	0.349	0.339	0.330	0.321	0.313	0.305	0.298	0.291	0.284
0.7	0.326	0.317	0.308	0.300	0.292	0.285	0.278	0.271	0.265	0.259
0.8	0.296	0.288	0.280	0.272	0.265	0.259	0.253	0.247	0.241	0.236
0.9	0.269	0.261	0.254	0.248	0.241	0.235	0.230	0.224	0.219	0.215
1.0	0.244	0.237	0.231	0.225	0.219	0.214	0.209	0.204	0.200	0.195
1.1	0.222	0.215	0.210	0.204	0.199	0.195	0.190	0.186	0.182	0.178
1.2	0.201	0.196	0.191	0.186	0.181	0.177	0.173	0.169	0.166	0.162
1.3	0.183	0.178	0.173	0.169	0.165	0.161	0.157	0.154	0.151	0.148
1.4	0.166	0.161	0.157	0.154	0.150	0.147	0.143	0.140	0.137	0.135
1.5	0.151	0.147	0.143	0.140	0.136	0.133	0.130	0.128	0.125	0.123
1.6	0.137	0.133	0.130	0.127	0.124	0.121	0.119	0.116	0.114	0.112
1.7	0.124	0.121	0.118	0.115	0.113	0.110	0.108	0.106	0.104	0.102
1.8	0.113	0.110	0.107	0.105	0.103	0.100	0.098	0.096	0.095	0.093
1.9	0.102	0.100	0.098	0.095	0.093	0.091	0.090	0.088	0.086	0.085
2.0	0.093	0.091	0.089	0.087	0.085	0.083	0.082	0.080	0.079	0.077
2.1	0.084	0.082	0.081	0.079	0.077	0.076	0.074	0.073	0.072	0.070
2.2	0.077	0.075	0.073	0.072	0.070	0.069	0.068	0.066	0.065	0.064
2.3	0.070	0.068	0.067	0.065	0.064	0.063	0.062	0.060	0.059	0.058
2.4	0.063	0.062	0.061	0.059	0.058	0.057	0.056	0.055	0.054	0.053
2.5	0.057	0.056	0.055	0.054	0.053	0.052	0.051	0.050	0.049	0.048
2.6	0.052	0.051	0.050	0.049	0.048	0.047	0.046	0.046	0.045	0.044
2.7	0.047	0.046	0.045	0.045	0.044	0.043	0.042	0.042	0.041	0.040
2.8	0.043	0.042	0.041	0.041	0.040	0.039	0.038	0.038	0.037	0.037
2.9	0.039	0.038	0.038	0.037	0.036	0.036	0.035	0.034	0.034	0.033
3.0	0.036	0.035	0.034	0.034	0.033	0.032	0.032	0.031	0.031	0.030

References

Aeronautical Radio, Inc., Reliability Research Dept., Washington, D.C., 1958, A selection of electron tube reliability functions, Publication No. 110.

Agree Report, 1957, Advisory Group on Reliability of Electronic Equipment, Reliability of Military Electronic Equipment, Office of the Assistant Secretary of Defense (Research and Engineering).

Arrow, K. J., S. Karlin, and H. Scarf, 1958, *Studies in the Mathematical Theory of Inventory and Production*, Stanford University Press, Stanford, Calif.

Barlow, R. E., 1962a, Repairman Problems, *Studies in Applied Probability and Management Science*, Chap. 2, edited by Arrow, Karlin, and Scarf, Stanford University Press, Stanford, Calif.

Barlow, R. E., 1962b, Applications of semi-Markov processes to counter problems, *Studies in Applied Probability and Management Science*, Chap. 3, edited by Arrow, Karlin, and Scarf, Stanford University Press, Stanford, Calif.

Barlow, R. E., and L. C. Hunter, 1960a, System efficiency and reliability, *Technometrics*, V. 2, No. 1, pp. 43–53.

Barlow, R. E., and L. C. Hunter, 1960b, Mathematical models for system reliability, *The Sylvania Technologist*, V. XIII, No. 1, 2.

Barlow, R. E., and L. C. Hunter, 1960c, Optimum preventive maintenance policies, *Operations Res.*, V. 8, No. 1, pp. 90–100.

Barlow, R. E., and L. C. Hunter, 1961, Reliability analysis of a one-unit system, *Operations Res.*, V. 9, No. 2, pp. 200–208.

Barlow, R. E., L. C. Hunter, and F. Proschan, 1963, Optimum checking procedures, *J. Soc. Indust. Appl. Math.*, V. 11, No. 4, pp. 1078–1095.

Barlow, R. E., L. C. Hunter, and F. Proschan, 1963, Optimum redundancy when components are subject to two kinds of failure, *J. Soc. Indust. Appl. Math.*, V. 11, No. 1, pp. 64–73.

Barlow, R. E., and A. W. Marshall, 1963, Tables of bounds for distributions with monotone hazard rate, Boeing Scientific Research Laboratories D1-82-0249.

Barlow, R. E., and A. W. Marshall, 1964, Bounds for distributions with monotone hazard rate, I and II, *Ann. Math. Statist.*, V. 35, No. 3, pp. 1234–1274.

Barlow, R. E., A. W. Marshall, and F. Proschan, 1963, Properties of probability distributions with monotone hazard rate, *Ann. Math. Statist*, V. 34, No. 2, pp. 375–389.

Barlow, R. E., and F. Proschan, 1962, Planned replacement, Chap. 4, *Studies in Applied Probability and Management Science*, edited by Arrow, Karlin, and Scarf, Stanford University Press, Stanford, Calif.

R. E. Barlow and F. Proschan, 1964, Comparison of replacement policies, and renewal theory implications, *Ann. Math. Statist.*, V. 35, No. 2, pp. 577–589.

Bellman, R., and S. Dreyfus, 1958, Dynamic programming and the reliability of multicomponent devices, *Operations Res.*, V. 6, No. 2, pp. 200–206.

Birnbaum, Z. W., 1955, On a use of the Mann-Whitney statistic, *Proceedings of the Third Berkeley Symposium on Mathematical Statistics and Probability*, University of California Press, Berkeley and Los Angeles, V. 1, pp. 13–17.

Birnbaum, Z. W., J. D. Esary, and S. C. Saunders, 1961, Multi-component systems and structures and their reliability, *Technometrics*, V. 3, No. 1, pp. 55–77.

Birnbaum, Z. W., and S. C. Saunders, 1958, A statistical model for life-length of materials, *J. Amer. Statist. Assoc.*, V. 53, No. 281, pp. 151–160.

Black, G., and F. Proschan, 1959, On optimal redundancy, *Operations Res.*, V. 7, No. 5, pp. 581–588.

Blackwell, D., 1962, Discrete dynamic programming, *Ann. Math. Statist.*, V. 33, No. 2, pp. 719–726.

Bourgin, D. G., 1939, Positive determinants, *Amer. Math. Monthly*, V. 46, pp. 225–226.

Brender, D. M., 1959, The statistical dynamics of preventive replacement, *Wescon Conv. Record*, pp. 23–36.

Brender, D. M., 1963, A surveillance model for recurrent events, IBM Watson Research Center report.

Buehler, R. J., 1957, Confidence intervals for the product of two binomial parameters, *J. Amer. Statist. Assoc.*, V. 52, pp. 482–493.

Campbell, N. R., 1941, The replacement of perishable members of a continually operating system, *J. Roy. Statist. Soc.*, V. 7, pp. 110–130.

Carhart, R. R., 1953, A survey of the current status of the electronic reliability problem, The RAND Corporation, ASTIA Document AD 80637.

Chung, K. L., 1960, *Markov chains with stationary transition probabilities*, Springer-Verlag, Berlin.

Clark, C. E., and G. T. Williams, 1958, Distributions of the members of an ordered sample, *Ann. Math. Statist.*, V. 29, No. 3, pp. 862–870.

Cox, D. R., 1963, *Renewal Theory*, John Wiley and Sons, New York.

Cox, D. R., and W. L. Smith, 1954, On the superposition of renewal processes, *Biometrika*, V. 41, pp. 91–99.

Cox, D. R., and W. L. Smith, 1961, *Queues*, Methuen Monographs, London.

Cramér, H., 1946, *Mathematical Methods of Statistics*, Princeton University Press, Princeton, N.J.

Creveling, C. J., 1956, Increasing the reliability of electronic equipment by the use of redundant circuits, *Proc. IRE*, V. 44, pp. 509–515.

Daniels, H. E., 1945, The statistical theory of the strength of bundles of threads, *Proc. Roy. Soc. London*, V. 183, pp. 405–435.

Dantzig, G. B., 1963, *Linear Programming and Extensions*, Princeton University Press, Princeton, N.J.

Davis, D. J., 1952, An analysis of some failure data, *J. Amer. Statist. Assoc.*, V. 47, No. 258, pp. 113–150.

Derman, C., 1961, On minimax surveillance schedules, *Naval Research Logistics Quarterly*, V. 8, No. 4, pp. 415–419.

Derman, C., 1962, On sequential decisions and Markov chains, *Management Science*, V. 9, No. 1, pp. 16–24.
Derman, C., 1963a, Stable sequential control rules and Markov chains, *J. Math. Analysis and Applications*, V. 6, No. 2, pp. 257–265.
Derman, C., 1963b, Optimal replacement and maintenance under Markovian deterioration with probability bounds on failure, *Management Science*, V. 9, No. 3, pp. 478–481.
Derman, C., 1963c, On optimal replacement rules when changes of state are Markovian, Chap. 9, *Mathematical Optimization Techniques*, edited by Richard Bellman, University of California Press, Berkeley and Los Angeles, pp. 201–210.
Derman, C., and J. Sacks, 1960, Replacement of periodically inspected equipment, *Naval Research Logistics Quarterly*, V. 7, No. 4, pp. 597–607.
Drenick, R. F., 1960a, Mathematical aspects of the reliability problem, *J. Soc. Indust. Appl. Math.*, V. 8, No. 1, pp. 125–149.
Drenick, R. F., 1960b, The failure law of complex equipment, *J. Soc. Indust. Appl. Math.*, V. 8, No. 4, pp. 680–690.
Duncan, A. J., 1953, *Quality Control and Industrial Statistics*, Richard D. Irwin, Homewood, Ill.
Epstein, B., 1948, Application of the theory of extreme values in fracture problems, *J. Amer. Statist. Assoc.*, V. 43, pp. 403–412.
Epstein, B., 1958, The exponential distribution and its role in life-testing, *Industrial Quality Control*. V. 15, No. 6, pp. 2–7.
Epstein, B., and M. Sobel, 1953, Life testing, *J. Amer. Statist. Assoc.*, V. 48, No. 263, pp. 486–502.
Esary, J. D., and F. Proschan, 1962, The reliability of coherent systems, *Redundancy Techniques for Computing Systems*, Spartan Books, Washington, D.C., pp. 47–61.
Esary, J. D., and F. Proschan, 1963a, Coherent structures of non-identical components, *Technometrics*, V. 5, No. 2, pp. 191–209.
Esary, J. D., and F. Proschan, 1963b, Relationship between system failure rate and component failure rate, *Technometrics*, V. 5, No. 2, pp. 183–189.
Everett, III, H., 1963, Generalized LaGrange multiplier method for solving problems of optimum allocation, of resources, *Operations Res.*, V. 11, pp. 399–417.
Fabens, A. J., 1961, The solution of queueing and inventory models by semi-Markov processes, *J. Roy. Statist. Soc.*, Series B, V. 23, No. 1, pp. 113–127.
Fan, K., and G. G. Lorentz, 1954, An integral inequality, *Amer. Math. Monthly*, V. 61, pp. 626–631.
Feller, W.,1941, On the integral equation of renewal theory, *Ann. Math. Statist.*, V. 12, pp. 243–267.
Feller, W., 1949, Fluctuation theory of recurrent events. *Trans. Amer. Math. Soc.*, V. 67, pp. 98–119.
Feller, W., 1957, *An Introduction to Probability Theory and Its Applications*, V. 1, 2nd ed., John Wiley and Sons, New York.
Feller, W., 1961, A simple proof for renewal theorems, *Comm. of Pure and Appl. Math.*, V. 14, pp. 285–293.
Flehinger, B. J., 1958, Reliability improvement through redundancy at various system levels, *IBM Journal of Research and Development*, V. 2, No. 2, pp. 148–158.
Flehinger, B. J., 1962a, A general model for the reliability analysis of systems under various preventive maintenance policies, *Ann. Math. Statist.*, V. 33, No. 1, pp. 137–156.
Flehinger, B. J., 1962b, A Markovian model for the analysis of the effects of marginal testing on system reliability, *Ann. Math. Statist.*, V. 33, No. 2, pp. 754–766.

Fréchet, M., 1950, *Recherches Théoriques Moderne sur le Calcul des Probabilitiés*, 2nd ed., V. 1, 2, Gauthier-Villars, Paris.

Freudenthal, A. M., 1960, Prediction of fatigue life, *J. Appl. Phys.*, V. 31, No. 12, pp. 2196–2198.

Gantmacher, F., and M. Krein, 1950, *Oscillation Matrices and Kernels and Small Vibrations of Mechanical Systems*, 2nd ed. (in Russian), State Publishing House for Technical-Theoretical Literature, Moscow (translated, Atomic Energy Commission, Oak Ridge, Tenn., 1961).

Geisler, M. A., and H. W. Karr, 1956, The design of military supply tables for spare parts, *Operations Res.*, V. 4, No. 4, pp. 431–442.

Gourary, M. H., 1956, An optimum allowance list model, *Naval Research Logistics Quarterly*, V. 3, No. 3, pp. 177–191.

Gourary, M. H., 1958, A simple rule for the consolidation of allowance lists, *Naval Research Logistics Quarterly*, V. 5, No. 1, pp. 1–15.

Gordon, R., 1957, Optimum component redundancy for maximum system reliability, *Operations Res.*, V. 5, No. 2, pp. 229–243.

Green, A. W., and R. C. Horne, 1961, Maintainability of shipboard electronic systems, Arinc Research Corporation, No. 118-4-228, Washington, D.C.

Gumbel, E. J., 1935, Les valeurs extremes des distributions statistiques, *Annales de l'Institute Henri Poincaré*, V. 4, Fasc. 2, p. 115.

Gumbel, E. J., 1958, *Statistics of Extremes*, Columbia University Press, New York.

Hanne, J. R., 1962, Optimizing simple circuitry for reliability and performance by failure mode, *Applications and Industry*, American Institute for Electrical Engineers.

Hardy, G. H., J. E. Littlewood, and G. Pólya, 1952, *Inequalities*, Cambridge University Press, New York.

Harris, T. E., 1952, First passage and recurrence distributions, *Trans. Amer. Math. Soc.*, V. 73, No. 3, pp. 471–486.

Herd, G. R., 1955, Failure rates, ARINC Monograph 2, Aeronautical Radio Inc., Washington, D.C.

Hirschmann, I. I., and D. V. Widder, 1955, *The Convolution Transform*, Princeton University Press, Princeton, N.J.

Hosford, J. E., 1960, Measures of dependability, *Operations Res.*, V. 8, No. 1, pp. 53–64.

Howard, R. A., 1960, *Dynamic Programming and Markov Processes*, M.I.T. Press.

Hunter, L. C., and F. Proschan, 1961, Replacement when constant failure rate precedes wearout, *Naval Research Logistics Quarterly*, V. 8, No. 2, pp. 127–136.

Jewell, W. S., 1963, Markov-renewal programming, I and II, *Operations Res.*, V. 11, No. 6, pp. 938–971.

Johnson, N. L., 1959, A proof of Wald's theorem on cumulative sums, *Ann. Math. Statist.*, V. 30, No. 4, pp. 1245–1247.

Jordan, C., 1950, *Calculus of Finite Differences*, Chelsea Publishing Co., New York.

Jorgenson, D., and R. Radner, 1962, Optimal replacement and inspection of stochastically failing equipment, Chap. 12, *Studies in Applied Probability and Management Science*, edited by Arrow, Karlin, and Scarf, Stanford University Press, Stanford, Calif.

Kao, J. H. K., 1956, A new life-quality measure for electron tubes, *IRE Transactions on Reliability and Quality Control*, PGRQC-7.

Kao, J. H. K., 1958, Computer methods for estimating Weibull parameters in reliability studies, *IRE Transactions on Reliability and Quality Control*, PGRQC-13, pp. 15–22.

Karlin, S., 1955, The structure of dynamic programming models, *Naval Research Logistics Quarterly*, V. 2, No. 4, pp. 285–294.

REFERENCES

Karlin, S., 1956, Decision theory for Pólya type distributions. Case of Two Actions, I, *Proceedings of the Third Berkeley Symposium on Mathematical Statistics and Probability*, University of California Press, Berkeley and Los Angeles, V. 1, pp. 115–128.

Karlin, S., 1957, Pólya-type distributions, II, *Ann. Math. Statist.*, V. 28, No. 2, pp. 281–308.

Karlin, S., 1958, Steady state solutions, Chap. 14, *Studies in the Mathematical Theory of Inventory and Production*, edited by Arrow, Karlin, and Scarf, Stanford University Press, Stanford, Calif., pp. 223–269.

Karlin, S., 1959, *Mathematical Methods and Theory in Games, Programming, and Economics*, Addison-Wesley Publishing Co., Cambridge, Mass.

Karlin, S., 1963, Total positivity and convexity preserving transformations, Proceedings of Symposia in Pure Mathematics, Vol. VII, *Convexity*, American Mathematical Society.

Karlin, S., 1964, Total positivity, absorption probabilities, and applications, *Trans. Amer. Math. Soc.*, V. 111, No. 1, pp. 33–107.

Karlin, S., (in preparation), *Total Positivity and Applications to Probability and Statistics*, Stanford University Press, Stanford, Calif.

Karlin, S., and J. McGregor, 1957, The classification of birth and death processes, *Trans. Amer. Math. Soc.*, V. 86, pp. 366–400.

Karlin, S., and J. McGregor, 1958, Linear growth, birth and death processes, *J. Math. and Mech.*, V. 7, No. 4, pp. 643–662.

Karlin, S., and J. McGregor, 1959a, Coincidence properties of birth and death processes, *Pacific J. Math.*, V. 9, No. 4, pp. 1109–1140.

Karlin, S., and J. McGregor, 1959b, Coincidence probabilities, *Pacific J. Math.*, V. 9, No. 4, pp. 1141–1164.

Karlin, S., and F. Proschan, 1960, Pólya type distribution of convolutions, *Ann. Math. Statist.*, V. 31, No. 3, pp. 721–736.

Karlin, S., F. Proschan, and R. E. Barlow, 1961, Moment inequalities of Pólya frequency functions, *Pacific J. Math.*, V. 11, No. 3, pp. 1023–1033.

Karlin, S., and H. Rubin, 1956, The theory of decision procedures for distribution with monotone likelihood ratio, *Ann. Math. Statist.*, V. 27, No. 2, pp. 272–299.

Kemeny, J., and J. Snell, 1960, *Finite Markov Chains*, D. Van Nostrand Co., Princeton, N.J.

Kendall, M. G., 1938, A new measure of rank correlation, *Biometrika*, V. 30, pp. 81–93.

Kettelle, J. D., Jr., 1962, Least-cost allocation of reliability investment, *Operations Res.*, V. 10, No. 2, pp. 249–265.

Khintchine, A. Ya., 1932, Mathematisches über die Erwartung von einen öffentlicher Schalter, *Matem. Sbornik*.

Khintchine, A. Ya., 1960, Mathematical methods in the theory of queueing, Chap. V, *Griffin's Statistical Monographs and Courses*, No. 7, edited by M. G. Kendall, London.

Klein, M., 1962, Inspection-maintenance-replacement schedules under Markovian deterioration, *Management Science*, V. 9, No. 1, pp. 25–32.

Kuhn, H. W., and A. W. Tucker, 1951, Nonlinear programming, *Proceedings of the Second Berkeley Symposium on Mathematical Statistics and Probability*, University of California Press, Berkeley and Los Angeles, pp. 481–492.

Lieblein, J., 1953, On the exact evaluation of the variances and covariances of order statistics in samples from the extreme value distribution, *Ann. Math. Statist.*, V. 24, No. 2, pp. 282–287.

Lieblein, J., and M. Zelen, 1956, Statistical investigation of the fatigue life of deep-groove ball bearings, *J. Res., Nat. Bureau Stand.*, V. 57, pp. 273–316.

Lotka, A. J., 1939, A contribution to the theory of self-renewing aggregates with special reference to industrial replacement, *Ann. Math. Statist.*, V. 10, pp. 1–25.

Madansky, A., 1958, Uses of tolerance limits in missile evaluation, *Proceedings of Statistical Techniques in Missile Evaluation Symposium*, Virginia Polytechnic Institute, Aug. 5–8.

Mann, H. B., 1945, Nonparametric tests against trend, *Econometrica*, V. 13, pp. 245–259.

Manne, A. S., 1960, Linear programming and sequential decisions, *Management Science* V. 6, No. 3, pp. 259–267.

Mine, H., 1959, Reliability of physical system, *Transactions of the 1959 International Symposium on Circuit and Information Theory*.

Molina, E. C., 1942, *Poisson's Exponential Binomial Limit*, D. Van Nostrand Co., Princeton, N.J.

Mood, A. M., 1950, *Introduction to the Theory of Statistics*, McGraw-Hill Book Co., New York.

Moore, E. F., and C. E. Shannon, 1956, Reliable circuits using less reliable relays, *J. of the Franklin Institute*, V. 262, Pt. I, pp. 191–208, and V. 262, Pt. II, pp. 281–297.

Morrison, D. F., and H. A. David, 1960, The life distribution and reliability of a system with spare components, *Ann. Math. Statist.*, V. 31, No. 4, pp. 1084–1094.

Morse, P. M., 1958, *Queues, Inventories, and Maintenance*, Chapter 11, John Wiley and Sons, New York.

Moskowitz, F., and J. B. McLean, 1956, Some reliability aspects of system design, *IRE Transactions on Reliability and Quality Control*, PGRQC-8, pp. 7–35.

National Bureau of Standards, 1950, *Tables of the Binomial Probability Distribution*, Applied Mathematics Series 6.

Ososkov, G. A., 1956, A limit theorem for flows of similar events, *Theory of Prob. and Its Appl.*, V. 1, pp. 248–255.

Palm, C., 1943, Intensitätsschwankungen im fernsprechverkehr, *Ericsson Technics*, No. 44, pp. 3–189.

Palm, C., 1947, Arbetskraftens Fordelning vid betjaning av automatskiner, *Industritidningen Norden*, 75.

Parzen, E., 1962, *Stochastic Processes*, Holden-Day, San Francisco, Calif.

Pearson, K., 1932, On the mean character and variance of a ranked individual, and on the mean and variance of the intervals between ranked individuals, *Biometrika*, V. 24, pp. 203–279.

Pearson, K., 1934, *Tables of the Incomplete Gamma Function*, Cambridge University Press, New York.

Pólya, G., and G. Szegö, 1925, *Aufgaben und Lehrsätze aus der Analysis*, V. I. Springer-Verlag, Berlin.

Product Engineering, 1960, *A Manual of Reliability*, McGraw-Hill Pub. Co., New York.

Proschan, F., 1960a, *Pólya Type Distributions in Renewal Theory, with an Application to an Inventory Problem*, Prentice-Hall, Englewood, N.J.

Proschan, F., 1960b, Optimal System Supply, *Naval Research Logistics Quarterly*, V. 7, No. 4, pp. 609–646.

Proschan, F., 1963, Theoretical explanation of observed decreasing failure rate, *Technometrics*, V. 5, No. 3, pp. 375–383.

Proschan, F., and T. A. Bray, 1963, Optimum redundancy under multiple constraints, Boeing Scientific Research Laboratories Document D1-82-0253.

Proschan, F., and R. Pyke (in preparation), Tests for increasing failure rate, Boeing Scientific Research Laboratories Document.

Pyke, R., 1961a, Markov renewal processes: Definitions and preliminary properties, *Ann. Math. Statist.*, V. 32, No. 4, pp. 1231–1242.

Pyke, R., 1961b, Markov renewal processes with finitely many states, *Ann. Math. Statist.*, V. 32, No. 4, pp. 1243–1259.

Radio-Electronics-Television Manufacturers Association, 1955, *Electronic Applications Reliability Review*, V. 3, No. 1, May, p. 18.

Radner, R., 1959, Limit distributions of failure time for series-parallel systems, Cowles Foundation Paper 139.

Radner, R., and D. W. Jorgenson, 1963, Opportunistic replacement of a single part in the presence of several monitored parts, *Management Science*, V. 10, No. 1, pp. 70–84.

Rényi, A., 1953, On the theory of order statistics, *Acta Math. Acad. Science Hungar.*, V. 4, pp. 191–231.

Roeloffs, R., 1962, Minimax surveillance schedules, Technical Report 16, Statistical Engineering Group, Columbia University, New York.

Rosenblatt, J. R., 1963, Confidence limits for the reliability of complex systems, *Statistical Theory of Reliability*, edited by M. Zelen, University of Wisconsin Press, Madison, Wis., pp. 115–148.

Rustagi, J. S., 1957, On minimizing and maximizing a certain integral with statistical applications, *Ann. Math. Statist.*, V. 28, pp. 309–328.

Savage, I. R., 1956, Cycling, *Naval Research Logistics Quarterly*, V. 3, No. 3, pp. 163–175.

Saunders, S. C., 1960, On classes of S-shaped functions, Boeing Scientific Research Laboratories Mathematical Note 216.

Schoenberg, I. J., 1951, On Pólya frequency functions I. The totally positive functions and their Laplace transforms, *Journal d'Analyse Mathématique*, Jerusalem, V. 1, pp. 331–374.

Schoenberg, I. J., 1953, On smoothing operations and their generating functions, *Bull. Amer. Math. Soc.*, V. 59, No. 3, pp. 199–230.

Sevastjanov, B. A., 1957, An ergodic theorem for Markov processes and its application to telephone systems with refusals, *Teor. Veroyatnost. i Primenen*, V. 2, pp. 106–116 (Russian, English summary).

Smith, W. L., 1954, Asymptotic renewal theorems, *Proc. Roy. Soc. Edinburgh*, Sect. A, V. 64, Pt. I (No. 2), pp. 9–48.

Smith, W. L., 1955, Regenerative stochastic processes, *Proc. Roy. Soc. London*, Ser. A, V. 232, pp. 6–31.

Smith, W. L., 1958, Renewal theory and its ramifications, *J. Roy. Statist. Soc.*, Ser. B, V. 20, No. 2, pp. 243–302.

Sobel, M., and J. A. Tischendorf, 1959, Acceptance sampling with new life test objectives, *Proceedings Fifth National Symposium on Reliability and Quality Control*, IRE, pp. 108–118.

Steck, G. P., 1957, Upper confidence limits for the failure probability of complex networks, *Sandia Corporation Research Report*, SC-4133(TR), Albuquerque, New Mexico.

Steffensen, J. F., 1950, *Some Recent Researches in the Theory of Statistics and Actuarial Science*, Cambridge University Press, New York.

Takács, L., 1951, Occurrence and coincidence phenomena in case of happenings with arbitrary distribution law of duration, *Acta Math. Acad. Sci. Hungar.*, pp. 275–298.

Takács, L., 1957a, On certain sojourn time problems in the theory of stochastic processes, *Acta Math. Acad. Sci. Hungar.*

Takács, L., 1957b, On a stochastic process concerning some waiting time problems, *Teor. Veroyatnost i Primenen*, V. 2, pp. 92–105.

Takács, L., 1958, On a combined waiting time and loss problem concerning telephone traffic, *Ann. Univ. Scient. Budapest*, Eotvos, Sect. Math. 1, pp. 73–82.

Takács, L., 1959, On a sojourn time problem in the theory of stochastic processes, *Trans. Amer. Math. Soc.*, V. 93, pp. 531–540.

Takács, L., 1962, *Introduction to the Theory of Queues*, Oxford University Press, New York.

Tate, R. F., 1959, Unbiased estimation: Functions of location and scale parameters, *Ann. Math. Statist.*, V. 30, No. 2, pp. 341–366.

Truelove, A. J., 1961, Strategic reliability and preventive maintenance, *Operations Res.*, V. 9, No. 1, pp. 22–29.

von Neumann, J., 1956, Probabilistic logics, *Automata Studies*, edited by C. E. Shannon and J. McCarthy, Princeton University Press, Princeton, N.J.

Weibull, W., 1939, A statistical theory of the strength of materials, *Ing. Vetenskaps Akad. Handl.*, No. 151.

Weibull, W., 1951, A statistical distribution function of wide applicability, *J. Appl. Mech.* 18, pp. 293–297.

Weiss, G., 1956a, On the theory of replacement of machinery with a random failure time, *Naval Research Logistics Quarterly*, V. 3, No. 4, pp. 279–293.

Weiss, G., 1956b, The reliability of a redundant system which operates repetitively, NAVORD 4348, Naval Ordnance Laboratory, White Oak, Maryland.

Weiss, G., 1956c, On some economic factors influencing a reliability program, NAVORD 4256, Naval Ordnance Laboratory, White Oak, Maryland.

Weiss, G., 1956d, A note on the coincidence of some random functions, *Quart. Appl. Math.*, V. 14, No. 1, pp. 103–107.

Welker, E. L., 1959, Relationship between equipment reliability, preventive maintenance policy, and operating costs, ARINC Monograph 7, Aeronautical Radio Inc., Washington, D.C.

Welker, E. L., and R. C. Horne, Jr., Concepts associated with system effectiveness, ARINC Monograph No. 9, Aeronautical Radio Inc., Washington, D.C.

Zahl, S., 1963, An allocation problem with applications to operations research and statistics, *Operations Res.*, V. 11, pp. 426–441.

Zelen, M., and M. C. Dannemiller, 1961, The robustness of life testing procedures derived from the exponential distribution, *Technometrics*, V. 3, No. 1, pp. 29–49.

Author Index

Arrow, K. J., 176, 227

Barlow, R. E., 7, 12, 22, 40, 57, 75, 81, 85, 107, 108, 163, 185
Bellman, R., 162
Birnbaum, Z. W., 2, 4, 5, 196
Black, G., 5
Blackwell, D., 156
Bray, T. A., 185
Brender, D. W., 47, 115
Buehler, R. J., 4

Campbell, N. R., 2, 27
Carhart, R. R., 3
Chung, K. L., 153
Clark, C. E., 195
Cox, D. R., 16, 21, 53, 139
Cramér, H., 176, 193

Daniels, H. E., 2
Dannemiller, M. C., 4
Dantzig, G. B., 158
David, H. A., 180
Davis, D. J., 4, 10
Derman, C., 86, 107, 114, 152, 153, 158
Doob, J. L., 55
Drenick, R. F., 8, 21, 67, 152, 162
Duncan, A. J., 2

Epstein, B., 2, 3, 13, 233
Erlang, A. K., 1
Esary, J. D., 5, 196, 202, 220
Everett, H., 162

Fabens, A. J., 136
Fan, K., 35
Feller, W., 2, 15, 51, 139

Flehinger, B. J., 47, 67, 74, 152
Fréchet, M., 129
Freudenthal, A. M., 11

Gantmacher, F., 227
Geisler, M. A., 162
Gordon, R., 163
Gourary, M. H., 162
Green, A. W., 11
Gumbel, E. J., 2, 10, 16
Gupta, S., 5

Hanne, J. R., 163
Hardy, G. H., 207
Harris, T. E., 144
Herd, G. R., 23, 47
Hirschmann, I. I., 227
Horne, R. C., 7, 11
Hosford, J. E., 7
Howard, R. A., 156
Hunter, L. C., 7, 47, 75, 88, 107, 108, 112, 163, 185

Jewell, W. S., 156
Johnson, N. L., 53
Jordan, C., 54
Jorgenson, D., 107, 117

Kao, J. H. K., 4, 16
Karlin, S., 25, 34, 61, 87, 117, 140, 142, 145, 154, 176, 225, 227, 229, 230, 235
Karr, H. W., 162
Kemeny, J., 122
Kendall, M. G., 233
Kettelle, J. D., 162, 170
Khintchine, A. Ya., 1, 20

AUTHOR'S INDEX

Klein, M., 119
Kolmogorov, B. V., 53
Krein, M., 227

Lieblein, J., 16, 194
Littlewood, J. E., 207
Lorentz, G. G., 35
Lotka, A. J., 2, 47
Lusser, R., 3

Madansky, A., 4
Mann, H. B., 233
Manne, A. S., 156
Marshall, A. W., 12, 22, 40, 45, 48
McGregor, J., 140, 142, 145, 154, 227
McLean, J. B., 162
Mendenhall, W., 5
Mihoc, 129
Mine, H., 162
Molina, E. C., 96, 177
Mood, A. M., 216
Moore, E. I., 4, 163, 196, 211
Morrison, D. F., 180
Morse, P. M., 86
Moskowitz, F., 162

Ososkov, G. A., 1, 20

Palm, C., 1
Parzen, E., 49, 57, 127
Pearson, K., 96
Pólya, G., 207, 228
Prokhorov, Yu. V., 53
Proschan, F., 5, 12, 22, 37, 48, 57, 61, 85, 107, 163, 169, 176, 185, 197, 202, 220, 225, 232
Pyke, R., 131, 232

Radner, R., 107
Rényi, A., 233
Roeloffs, R., 114
Rosenblatt, J. R., 4
Rubin, H., 25, 227
Rustagi, J. S., 34

Sacks, J., 86
Saunders, S. C., 2, 196
Savage, I. R., 95, 107
Scarf, H., 176, 227
Schoenberg, I. J., 228, 229
Sevastjanov, B. A., 148
Shannon, C. E., 4, 163, 196
Smith, W. L., 4, 21, 48, 56, 60, 136, 139
Snell, J., 122
Sobel, M., 3, 4, 5, 233
Steck, G. P., 4
Steffenson, J. F., 10
Szegö, G., 228

Takács, L., 47, 59, 75, 79, 139, 147, 150
Tate, R. F., 5
Tischendorf, J. A., 5
Truelove, A. J., 8

von Neumann, J., 4, 196

Weibull, W., 2, 16
Weiss, G., 4, 47, 62, 75, 90
Welker, E. L., 7, 67
Widder, D. V., 227
Williams, G. T., 195

Zahl, S., 162
Zelen, M., 4, 16

Subject Index

Absolute continuity of an IFR distribution, 26
Absorbing Markov chain, 125
Absorbing semi-Markov process, 135
Absorbing state, 123
Age replacement, 46, 61, 67, 85
   comparison with block replacement, 67–72
   existence of optimum policy, 92
   minimax strategy, 91–92
AGREE report, 3, 4
Allocation, 60, 162, 163
   complete family of undominated allocations, 180–185
   of redundancy, 50, 164
ARINC, 3
Asymptotic relative efficiency, 234
Availability, 5, 7
   limiting availability, 85

Binomial distribution, 17–18
Birth and death process, 140, 142, 154
Blackwell's theorem, 51, 55, 59
Block replacement, 46, 67, 85, 95
   comparison with age replacement, 67–72
Bounds on density, 42
Bounds on DFR distribution, 27, 28, 29, 31, 32, 40, 239–240
Bounds on discrete distribution, 43
Bounds on IFR distribution, 27, 28, 29, 40, 41, 236–238
Bounds on percentiles, 30
Bounds on renewal function, 52, 53
Bounds on structure failure rate, 223–226

Bridge network, 204
Burn-in test, 10

Chebyshev bounds, 12
Chebyshev inequality, 207
Checking (*see* Inspection)
Classification of states for a Markov chain, 123
Coefficient of variation, 235
   of DFR distribution, 33
   of IFR distribution, 33
Coherent structure, 197, 203
Communicating states, 123
Composition of structures, 200, 201
Control limit rule, 152, 154, 155
Convolution
   of decreasing failure rate distributions, 37
   of increasing failure rate distributions, 36–37
Cut, 203, 209
Cyclic rule, 160

Death process, 155
Debugging, 22
Decreasing failure rate (DFR), 10, 11, 23, 25
   and replacement, 85
   bounds on survival probability, 12, 31, 40, 41, 239–240
   closed under mixture, 37
   comparison with exponential, 26–39
   definition of, 12
   test of, 12
Decreasing failure rate distribution
   absolute continuity of, 26

251

Decreasing failure rate distribution, coefficient of variation of, 33
  comparison with exponential distribution, 26–35
  convolution of, 37
  log convexity of, 25
  mixture of, 37
  moments of, 32, 33
Decreasing mean residual life, 11, 43
  and moments, 43
Density, bounds on, 45
Diffusion process, 145
Discrete failure distribution, 17, 18
Discrete failure rate, 43
Discrete IFR distribution, 24
Discrete time Markov chain, 122
Down time, 79–80, 85
  distribution of, 79
Dual to failure rate, 37, 38
Dynamic programming, 117, 162

Early failure, 9, 10
Effectiveness, 5
Efficiency, 5, 7
  limiting, 8
Ehrenfest model, 145
Elementary renewal theorem, 55
Embedded Markov chain, 123, 144
Ergodic Markov chain, 127
Ergodic states, 123
Erlang's formula, 148
Essential component, 210
Essentiality, 162
Extreme value distribution, 22
Excess random variable, 51
  distribution for, 58
Expected gain, 5, 6, 7
Expected number of removals, 66
Exponential distribution, 1, 3, 4, 5, 10, 11, 15, 16, 176, 192, 222
  and Markov process, 119
  and redundancy, 192–193
  as failure law of complex equipment, 18–22
  comparison with DFR, 26–39
  comparison with IFR, 26–39
  definition of, 13
  failure rate of, 13
  mixture of, 37

Exponential distribution, spares for, 176–180
  two parameter, 5
Extreme value distribution, 2, 10, 13, 16, 17
  definition of, 13
  failure rate of, 13

Fail-safe structure, 197, 216
Failure distributions, 9–45
  discrete, 17
Failure rate, 4
  bounds on, 45
  definition of, 22
  general, 41–45
  instantaneous, 23
  monotone, 22–35
  of binomial distribution, 18
  of convolution, 38
  of exponential distribution, 13
  of extreme value distribution, 13
  of gamma distribution, 13, 14
  of geometric distribution, 18
  of log normal distribution, 13
  of negative binomial distribution, 18
  of normal distribution, 13, 15
  of Poisson distribution, 18
  of system, 38
  of Weibull distribution, 13, 14
Failure rate function, 10
  of discrete distribution, 18
Fatigue in materials, 2
Finite time span replacement, 85
  age replacement, 92–93
First-passage time, 130, 132–134, 142
Force of mortality, 10
Fundamental renewal equation, 50
Fundamental renewal theorem, 59

Gamma distribution, 2, 5, 9, 11, 13, 14, 16, 234
  and spares, 180
  failure rate of, 13, 14
Generalized renewal quantity, 135
Geometric distribution, 17
  definition of, 17
  failure rate of, 18
Guarantee time, 13

SUBJECT INDEX

Hazard rate, 10, 23

IFR Markov chain, 121, 154
Increasing failure rate (IFR), 11, 23, 25, 214, 215, 219, 222
  and convolution, 36
  bounds on survival probability, 12, 28, 29, 41, 236–238
  closed under convolution, 36
  comparison with exponential, 26–39
  definition of, 12, 23
  test for, 12, 232–235
Increasing failure rate distribution
  and absolute continuity, 26
  bounds on mean, 30
  bounds on percentile, 30
  coefficient of variation of, 33
  comparison with exponential distribution, 26–35
  definition of discrete IFR distribution, 24
  log concavity of, 25
  mixture, 37
  moments of, 32, 33
  order statistics of, 36, 38
  tables of bounds, 236–240
Infinite time span replacement, 85
  age replacement, 86–92
Inspection policies, 107–117
  for normal distribution, 113
  for Pólya frequency function, 110–111
  for uniform distribution, 113
  minimax, 107, 114
  optimum, 107
  with renewal, 114–117
Instantaneous failure rate, 23
Intensity function, 10
Interval availability, 5
  limiting, 7, 8
Interval reliability, 8, 74, 82
  limiting, 74
Irreducible Markov chain, 160

Jensen's inequality, 27

Kettelle's algorithm, 172, 173, 180–185
k-out-of-n structures, 36, 197, 213, 216–219, 221, 222, 224

Life testing, 2, 3, 5, 30
Likelihood ratio test, 234
Limiting efficiency, 8
Limiting interval availability, 7, 8
Linear programming, 121, 153, 156, 158, 161
Log normal distribution, 9, 11, 13, 17
Logistic problems, 5

Maintainability, 4
Maintenance policies, 1
  operating characteristics of, 46–83
  opportunistic, 107, 117–118
  optimum, 84
Maintenance ratio, 8
Marginal checking, 121, 151–156
Marginal checking intervals, 155
Marginal states, 152, 155
Markov bound, 236
Markov chain, 121–131, 152, 157
  absorbing, 125
  classification of states, 123
  discrete time, 122
  embedded, 123, 144
  ergodic, 123
  IFR Markov chain, 121, 154
  irreducible, 160
  time homogeneous, 131
  transition matrix, 122
Markovian deterioration, 156–161
Markov process, 119
  and exponential distribution, 119
  definition of, 121
  stationary, 122
Markov renewal process, 131, 156
Mean residual life, 11, 47, 53
  and replacement policy, 47
Mean time to failure with replacement, 61
Mean time to in-service failure
  bounds on, 62
Median, 31, 63
Mill's ratio, 10
Minimax checking, 114
Minimax inspection schedule, 107, 114
Mission success ratio, 8
Mission survival probability, 8
Moment inequalities, 33, 43

SUBJECT INDEX

Moments of increasing failure rate distribution, 32, 33
  and decreasing mean residual life, 43
Monotone failure rate, 22–35, 227, 229
  preservation of, 35–39
Monotonic structure, 197, 202–226
Moore-Shannon inequality, 199, 211, 213, 216
Multicomponent structure, 196–225

Negative binomial distribution, 17
  definition of, 17
  failure rate of, 18
Nonperiodic Markov chain, 123, 129
Normal distribution, 2, 11, 16
  failure rate of, 13, 15
  inspection schedule for, 113
  truncated, 13, 88
  truncated, definition of, 13

Open-circuit failure, 163, 186, 197
Operating characteristics of maintenance policy, 46–83
Operating time, 80, 81
Opportunistic replacement, 117
Optimum inspection schedule, 107
Order statistics, 38, 221
  of increasing failure rate distribution, 36, 38

Parallel redundancy, 162, 164–165, 170–175
Parallel-series, 186
Parallel system, 28, 33, 39, 63, 65
Pascal distribution, 17
Path, 203, 209
Path and cut bounds on structure reliability, 206–209
Percentile, 27
Periodic replacement, 84, 85, 96
  with minimal repair, 96–98
Period of state, 123
Planned replacement, 85
Pointwise availability, 7
Poisson distribution, 1, 17, 49
  definition of, 17
  failure rate of, 18
Poisson process, 10, 11, 15, 19, 20, 51, 56, 57

Poisson process, expected number of renewals in, 51
Pólya frequency density of order two, 60, 61
  and inspection schedule, 110–111
Pólya frequency function of order two, 24, 110, 228, 229
Pólya frequency sequence of order two, 228
Pooled samples, 37
Preparedness models, 84, 107

Quality control, 2
Queueing, 139
Quorum function, 199, 218

Random age replacement, 46, 72–74, 85–95
Random walk, 144
Randomized stable rule, 156, 157, 160
Redundancy, 139
  allocation of, 164
  for exponential failure distribution, 192–193
  for uniform failure distribution, 193–194
  optimization, 162–195
  parallel, 162, 164–165, 170–175
  standby, 162, 165, 175–180
  two types of failure, 185–195
  undominated allocation, 164–166
Redundancy allocations
  complete family of, 166
  procedure 1, 166, 173
  procedure 2, 169, 174
Redundant structures
  replacement of, 63
Relay network, 163, 186, 196, 197–202, 209
Reliability, 209
  bounds on structure, 206
  definitions of, 5, 6, 7
  function, 205
  maximizing system, 186
Reliability function of structure, 204
Removals, 61
  expected number of, 66
  time between, 66
Renewal counting process, 48, 53

SUBJECT INDEX   255

Renewal density, 50
  bounds on, 59
  for gamma density, 57
Renewal density theorem, 60
Renewal function,
  bounds on, 53, 54, 68, 69, 71, 72
  definition of, 50
  for shifted exponential, 57
  for truncated exponential, 57
Renewal process, 8, 19
  definition of, 48
  underlying IFR distribution, 49
Renewal quantity
  generalized, 135
Renewal random variable, 48
  bounds on moments of, 54
  distribution for gamma density, 56
  positivity properties of, 60
Renewal theory, 2, 4, 48
Repairman problems, 121, 139–151
Repair time distribution, 17
Replacement, 2, 5
  age, 61–66, 85
  age, for finite time span, 92–95
  age, for infinite time span, 86–92
  and DFR distributions, 85
  block, 85
  bound on time to failure, 62
  existence of optimum age policy, 86, 92
  expected number of failures, 66
  exponential bounds on operating characteristics, 65, 66
  finite time span, 85
  for truncated normal, 88–91
  infinite time span, 85
  mean time to failure, 61
  minimax strategy for age, 91, 92
  of redundant structures, 63–65
  opportunistic, 117
  optimum age, 85–95
  optimum block, 95–96
  optimum block, for gamma, 95–96
  periodic, 84, 85, 96
  periodic, with minimal repair, 96–98
  planned, 85
  random age, 72–74, 86
  sequential, 84, 85, 98–107

Replacement, sequential, for gamma distribution, 104–107
  underlying IFR failure distribution, 62
Replacement rate, 8
Replacement rule, 153, 154
Reverse regular function, 228

Semi-Markov process, 4, 121, 131–139
  absorbing, 135
  limit theorems, 135–137
Sequential replacement, 84, 85, 98–107
  for gamma distribution, 104–107
Series-parallel, 186
Series system, 28, 33, 39, 63, 162, 223
Series system of IFR (DFR) components, 33–34
Servicing reliability, 83
Sheffer stroke organ, 196
Shortage, 162
Shortage random variable, 51, 58
  distribution for, 58
Short-circuit failure, 163, 186, 197
Sign reverse regular, 228
Single unit
  failure and repair, 74–77
  exponential failure and constant repair, 78–79
  exponential failure and repair, 77–78, 80
Size of path (cut), 204
Spare parts problem, 49
Spares, 36, 38, 49, 139, 162, 165
  for exponential failure distribution, 176–180
  for gamma failure distribution, 180
S-shapedness property, 196, 197, 209–216
Standby redundancy, 162, 165, 175–180
States
  absorbing, 123
  communicating, 123
  ergodic, 123
  period of, 123
Stationary Markov process, 122
Strategic reliability, 8
Structure failure rate, and component failure rates, 219–226

Structure failure rate, and component failure rates, bounds on, 223
Structure function, 202
Superposition of renewal processes, 19
Survival probability, under age replacement, 61
System survival probability, bounds for, 28

Tables of bounds for IFR (DFR) distributions, 236–240
Telephone trunking, 1, 11, 148, 149
Test for increasing failure rate, 232–235
Time between removals, 66
Time homogeneous Markov chain, 131
Totally positive functions, 25
Total positivity, 5, 227–231
  definition of, 227
Tradeoff between redundancy and replacement, 64
Transient state, 123

Two parameter exponential distribution, 5
Two-terminal network, 203, 208

Unbiased test, 233
Undominated redundancy allocation, 164–166
Uniform distribution, 113, 193
  inspection schedule for, 113
  redundancy, 193–194
Unimodality, 230

Variation diminishing property, 228, 229

Weibull distribution, 2, 4, 5, 9, 11, 13, 14, 16, 98, 194, 234
  definition of, 13
  failure rate of, 13, 14
Work hardening, 10